GARDENING *with*

HEIRLOOM PLANTS

DAVID
STUART

The Reader's Digest Association, Inc.
Pleasantville, New York/Montreal

A Reader's Digest Book

First published in Great Britain in 1997 by
Conran Octopus Limited as *Gardening with Antique Plants*

COMMISSIONING EDITOR Stuart Cooper
PROJECT EDITOR Caroline Taggart
ART EDITOR Alison Barclay
DESIGNER Isabel de Cordova
PICTURE RESEARCH Julia Pashley
PRODUCTION Julian Deeming
INDEX Margaret Pearcy

Library of Congress Cataloging in Publication Data
has been applied for.

ISBN 0-7621-0001-X

Printed in China

CONTENTS

The Story of Heirloom Plants

In 1896, an American writer, George H. Ellwanger, wrote ecstatically of his own, and neighboring, gardens in Rochester, New York: "You would know by the scent of the lilies that summer was here. How fragrant the censer of June! How profuse the scent of blossoming vegetation! — odors not alone from myriads of plants, but breathing from orchards, hedges, and thickets, rising from woods and hillsides, blown from meadows and pastures..." He admired, too, the gardens of "tumble down farmsteads" where the loveliest American flowers and the loveliest of Europe, too, filled the air with perfume. But he, like many of his European counterparts, was beginning to regret the loss of the very flowers of those old farmstead gardens, all being speedily replaced by new discoveries and the then-new "bedding" style of gardening. He had a long list of plants that he sorely missed: snowdrops, the old double daffodils, imperials (he meant *Fritillaria imperialis*), muscaris, larkspurs, campanulas of all sorts, bachelors' buttons, monkshoods, double white poppies, sweet clover, snow pink, white phloxes, dicentras, sweet Williams (grown in Europe since the sixteenth century), tall yellow tulips, tradescantias, sweet peas, valerians, madonna lilies (the same as those grown in the most ancient gardens of Mesopotamia thousands of years ago), white and purple stocks, lily-of-the-valley, briar rose (which the first Dutch settlers brought to Manhattan), white daylily (he probably meant hosta), tiger lilies, dahlias, hollyhocks (grown in New England by 1621), and even sunflowers.

This book is the story of these and many other antique plants, of their origins, of how they have been used in gardens, and how the many that can still be grown may be used today. It covers all sorts of species, from wildflowers native to North America or Europe to plants gathered in some foreign clime and grown here as exotics.

Opposite: *Canarina canariense*, the Canary Island bellflower. This one, painted for *Botanical Magazine* in the late eighteenth century, still makes a charming perennial climber in frostfree gardens.

Left: At the height of their popularity in the eighteenth century, uniformly colored tulips were called "mothers"; the color breaks that transformed them into gorgeous "flamed" or "feathered" sorts were caused by a virus that affected different layers of the petals in different ways.

Above: *Kennedia coccinea*, painted for *Botanical Magazine*, can make a spectacular antique climber for warm gardens or solaria.

WHAT IS AN HEIRLOOM?

In gardening terms, an "heirloom" or "antique" plant is anything more than a hundred years old. This is a tiny bit of time for plants in evolutionary terms, while it is an age for garden flowers or fruits, bred as the latest thing, the most brilliant, or the highest yielding. It is long enough to see them go out of fashion, and then become extinct.

In most important groups of garden plants such as roses or apples, carnations or crocuses, new varieties have continued to appear throughout the garden history of the species. You will find in these pages plant varieties that are still commercially available, but that relate to a specific epoch in the garden. In some genera (apples, roses, grapevines, irises, tulips, and so on, each of which easily merits a book to itself), there are still huge numbers of varieties around. While not all are antique, and fewer still are "deep" antiques from the very beginning of gardening, the treatment here is necessarily brief; I have chosen mainly species and varieties that I have grown and liked and that will associate well with other plants in this book.

Of course, almost every pure plant species is an heirloom, and what grows in the garden now is exactly the same as what might have grown in medieval gardens, or in the wilds of Afghanistan, or on the banks of the Mekong River long ages before gardening even began. After all, the species might be tens of thousands, or even millions of years old. For the purposes of this book, wild species are "aged" by the date when they were first cultivated in the documented garden; so, for instance, many orchid species, having grown for millions of years among the damp branches of Amazonian jungles, are here called "nineteenth-century," because that is when they began to be grown in European and American greenhouses — even though some of them may, perhaps much earlier, have been used by Native Americans.

But are there really many heirloom plants left? The answer is yes; there are some astonishing, and ancient, survivals. Your garden, whatever its size, can grow at least some of the flowers found in Dutch flower paintings of the seventeenth century, tulips from sixteenth-century Constantinople; or the enchanting double-flowered chelidonium found in a hedge by the botanist John Rea in the 1630s. You can still find the strange little "Plymouth strawberry," discovered in that English town in 1632; it was worn on ladies' bodices for decoration and is so strange that it has intrigued just enough gardeners to keep it alive. You can grow the Maltese-Cross (*Lychnis chalcedonica*), probably brought to western European gardens by returning Crusaders, or have roses from ancient Rome, or squashes that were known in Central America in the fifteenth century.

There are certainly plenty of plants to make an enchanting heirloom garden. The aim of this book is to help you find some of the loveliest heirlooms, and to show how they can be grown in contemporary gardens. By that, I mean gardens that are not museum pieces, but which are, nevertheless, filled with the beauty and romance of ancient plants.

Below: A Victorian seed catalog from Benary, a German firm that still exists, showing contemporary varieties of garden pea.

Right: In Europe, prize-winning gooseberries, often bred by artisan enthusiasts, reached prodigious sizes; the tiny leaves shown here may reflect the scale of the fruit. Many varieties with these colors and textures can still be grown.

Opposite: A rose from the hand of Redouté, who painted many roses for the Empress Josephine in the early nineteenth century. Though this variety may be lost, 'Tuscany Superb' makes a close and easily grown substitute.

THE ROMANCE OF HEIRLOOM PLANTS

For me, the real fascination of many heirloom plants is the way in which so many are so intimately bound up with the history of conquests and wars, of trade, of peace and plenty, and of social change. It's hard not to find a bowl of cherries from your own garden more romantic when you know how much they were valued by Mithridates, King of Pontus, a ruler so proud of his great knowledge that he wrote one of the first plant treatises, in the first century B.C. When his kingdom was destroyed by the Roman general Lucullus, the cherries, unknown to the Romans, were carried back to Italy, and the trees were among the treasures displayed at Lucullus's subsequent "triumph."

By the same token, it is difficult, when planting a tulip bulb, not to think of the stories of their introduction to the West. In one version, an ambassador at the court of Suleiman the Magnificent is supposed to have sent seeds back to Vienna in 1554. The first tulip from this seed flowered in a private garden in Augsburg in 1559.

Another version of the story concerns a cargo of Turkish bulbs that apparently arrived in Antwerp in 1562. The burghers of that great city, not knowing what the bulbs were, tried cooking them. Finding them not to their taste, they threw the remaining uncooked ones onto a midden, where a few of them subsequently flowered and caused a storm of interest. If this story is true, however, the bulbs must have been wild species or unfashionable varieties, for choice Turkish garden tulips were far too expensive in Constantinople to be bought by the sackful, even by the rich merchants of Antwerp.

Even that modern staple of late summer cottage gardens, the Japanese anemone (*Anemone hupehensis*), has drama; the form with purple semidouble flowers was discovered in 1844 growing on tombs on the ramparts of Shanghai, and the great collector Robert Fortune almost lost his life to pirates as he attempted to bring the plants home.

FLORISTS AND COLLECTORS

Plant stories are also bound up with the stories of ordinary and forgotten gardeners, from one Mistress Tuggy, who in Elizabethan London was following her late husband's example by selecting auricula seedlings so pretty that she gets a mention in John Parkinson's great garden book of 1629, *Paradisi in Sole Paradisus Terrestris*, to another of Parkinson's contacts, "my good friend Vincent Sion." None of Mistress Tuggy's auriculas seem to have survived, whereas the old double narcissus called 'Van Sion,' after Vincent, is still delighting admirers of daffodils.

These two gardeners were among the first "florists"; not florists in the way that we now use the word — of people who sell flowers — but gardeners obsessed enough with their favorite types of plant to sow endless seed, nurture the seedlings, grow them until they flowered, and then save but two or three of the most enchanting of the results. Some well-documented florists are still memorialized in plant names, like the ones called 'Sam Barlow.' Mr. Barlow (1824–88), from modest beginnings, became a partner in a bleach works at Rochdale in Lancashire, England. Selecting new forms of flowers that would win prizes on the show bench was his hobby, and he was so good at it that he became known as "King of the Northern Florists," particularly for his auriculas: he grew 500 seedlings every season, keeping only the two or three most beautiful. He has a gorgeous florists' tulip named after him (not, however, of his breeding, and not in commerce) and a garden pink, maroon centered and smelling of cloves and lemon.

Mr. Barlow, who was educated and eventually became prosperous, was among the "stars" of the day. The push forward in the creation of new varieties, or even of whole groups of varieties, sometimes came from wealthy collectors or, in the late nineteenth century, from big nurseries. But more often it came from unnamed and now totally forgotten individual "florists." Most of them were members of the manufacturing classes, usually urban; they grew the plants for showing at local meetings and took the best plants of all to larger shows in the big cities. The rewards were not just aesthetic; good "show" plants won money and found a ready market for cuttings and offsets.

Almost any plant in the garden could be adopted as a florists' flower, though obviously ones that held a lot of genetic diversity, or that could be crossed with related species, were the best choices. Important antique florists' groups included auriculas, violas, tulips, roses, hyacinths, anemones. Modern equivalents are irises, daylilies, ericas, and so on. But genetic diversity was apparent in fruit, too, so some easily grown fruits were bred in backyards and had clubs devoted to showing the results of breeding: the gooseberry was a popular "subject."

However, as with modern intensely bred groups, what made the difference between new "varieties" became smaller and smaller, until the varieties were scarcely distinguishable. As now, the fashion for a particular group lost momentum, and the florists' attention then shifted elsewhere in pursuit of some other ideal. Huge numbers of plants were thrown out, and the ones that somehow survived were confused with others that looked almost the same. Even those became infected with viruses, lost vigor and their names, and retreated into obscurity with their breeders. Some old florists' societies still exist, from the tiny and sadly diminishing Paisley Florists' Club in Scotland (founded in 1795 for the showing of anemones, pinks, carnations, hyacinths, and tulips, though now specializing in overweight vegetables) to the Wakefield Tulip Society in England, formed probably in the early nineteenth century and still having annual shows of old "flamed" tulips.

Opposite: Nineteenth-century variants of Swiss chard, some with deep ruby stems, others with "savoyed" foliage; comparable forms can still be found.

Below: Although the name of this fine narcissus (painted by Redouté in the nineteenth century) can no longer be traced, similar dark-cupped varieties are still grown.

Gardeners at the other end of the social scale became equally passionate about plant variation and the creation of new varieties; the aristocratic Thomas Knight — brother and eventual heir to the rakish Richard, whose obsessive interest in classical civilization made him a leading propagandist for the nineteenth-century developments in the landscape style, of which more later — spent his time (and money) quietly at Downton Castle in Shropshire, writing about horticulture and breeding new strawberries from the exciting American species arriving in Europe. He was responsible for huge numbers of publications, all of which survive. None of his strawberries do.

But florists were gardeners who did not stray beyond their own fences. Some garden plants, however, demanded much more adventurousness from their devotees, and risk was had in plenty by some of the great plant hunters, from the Tradescants, father and son, to the nineteenth-century collectors like David Douglas and Robert Fortune. John Tradescant the Elder (1570–1638) journeyed to Europe several times, even visiting the island of Cos, from which he collected the first 'Cos' lettuces to be seen in Europe. Although Cos lettuces originated in ancient Egypt, Tradescant's name for them has stuck. He went on to explore the farthest points of Russia, from where he brought back the first larch tree to be seen in western Europe. John the Younger (1608–62) was even more daring, making the trip to North America three, perhaps four times, and bringing back the first American asters.

However, the Golden Age of collecting from the Americas was the early decades of the nineteenth century, and perhaps the most influential collector was David Douglas. Financed by what became the Royal Horticultural Society, he explored the then unimaginably remote west coast of America, the Rockies, and the central states. Often enduring appalling conditions, he collected immensely important plants, from the tiny *Penstemon scouleri* to the vast *Sequoia sempervirens*, and including camassias, flowering currants, *Garrya elliptica,* and many more conifers, including the Douglas fir (*Pseudotsuga menziesii*). On a final trip, he visited Hawaii. He died there and was found, mangled, at the bottom of a staked pit that should have caught wild cattle. His last collections were beside him.

In the East, trampled almost bare by the number of collectors looking for rhododendrons, roses, and rare alpines, one of the most notable collectors was Robert Fortune. By 1847 he was busily buying up all the interesting camellias (as well as moutan peonies and dozens of other good plants) from Chinese nurseries, private gardens, and wild hillsides. Like David Douglas and many other collectors (and indeed, many head gardeners of the time), Fortune was a Scot, born in Berwickshire in 1812. He eventually became a gardener at the Edinburgh Botanic Garden; from there he moved to Chiswick House, in west London, where he looked after the tropical collection. While he was in the capital, China was opened to foreigners, and following the Treaty of Nanking in 1842, Fortune was sent out to find as many good garden plants as he could.

For a man who had never traveled out of Britain and who had hated the sea passage from Leith to London, this must have been a daunting prospect. It was nevertheless a dazzling success. In spite of thieves, beggars, con men, pirates, obstructive mandarins, and brushes with various oriental diseases, Fortune finally returned to London loaded with gorgeous plants. Many of them were brought back in Wardian cases (*see p.149*), and many plants that would otherwise not have survived the long and arduous journey were soon in full bloom at Chiswick. Fortune eventually retired, wrote bestselling memoirs, and died at home.

Opposite: Hyacinths with dense heads of flowers were favorites in Victorian flower beds, and were bred to make intense masses of color. Older varieties were often much more lax.

FINDING OLD FLOWERS

However exciting it is to gather a collection of antique flowers and make a garden with them, even more exciting is the hunt for forgotten flowers still hanging on in cottage gardens or abandoned patches by the wayside. But don't underestimate the difficulties. The chances of finding a lovely old plant, whether daylily, carnation, or apple tree, still with a moldering (and decipherable) label attached to it are minute.

To make matters worse, it is very hard to identify old garden flowers from original sources. Descriptions in eighteenth-century or earlier garden books are almost never adequately detailed, or well enough illustrated, to allow precise identification. Woodcuts (when available) are some help, but not much, and herbarium specimens are virtually useless — every herbarium seems to have sheets of brown pressed tulips, brown carnations, and brown auriculas. Though it is fascinating to see late seventeenth-century garden flowers in the flesh, none of the specimens is ever much use in matching with a living plant that the eager gardener hopes to clinch as an ancient survivor.

The most that can be said for some garden flowers, including a number in this book, is that they clearly belong to old "groups," say the "pheasant's eye" pinks, the "rennet" apples, or the "tazetta" narcissus. They can never be referred to with any certainty by an old variety name. Indeed, the plant may not have had one, if it was just a local variant or never in commerce. For most gardeners, perhaps, that doesn't matter; as long as a plant clearly belongs to an antique group, then it should be welcomed into the garden. There have, though, been attempts at crossing some of the best antiques to give them modern attributes, as with the roses that almost look like old ones but flower all year, or laced pinks that do the same. Such plants have their advantages, but they have almost always also lost much of their charm.

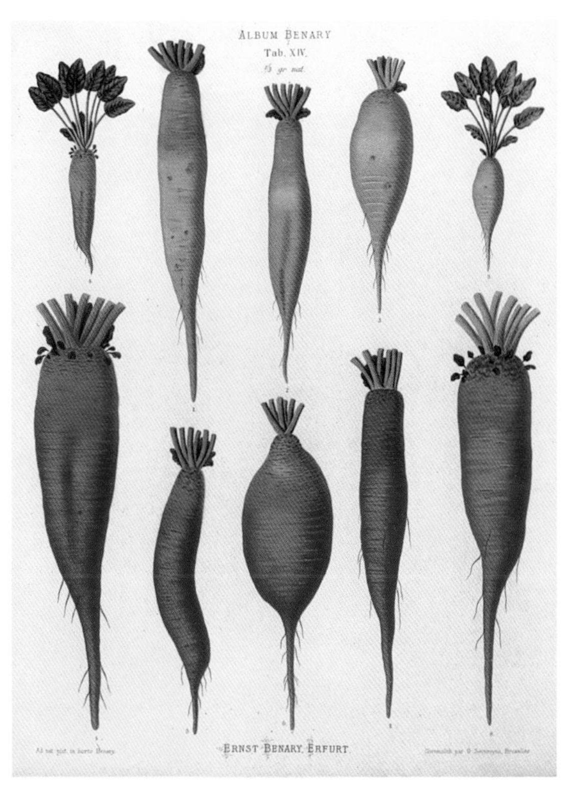

Opposite: The 1830 catalog of the London nursery Loddiges lists 40 species of *Dianthus*, without even beginning on the immense numbers of cultivated varieties.

Left: More plates from the Benary catalog; many of the squash varieties can still be found and can be grown in most gardens.

GROWING HEIRLOOM PLANTS

Although we can pinpoint the date when many plants appeared in gardens, or garden literature, for the first time, none need be rigorously related to a particular place or historical time. The deliberate use of antique plants is a recent phenomenon that doesn't need to be treated as an academic exercise; the plants can be grown for your own pleasure in a garden with no past or in a single pot on the windowsill. They are for everyone, not just the owners of historical acres or grand houses. This book suggests ways of growing them so that they will look beautiful, crop well, and smell delicious.

They can also lead on to other interests that can help in the search for antique plants, whether that is to visit other gardens or the collections of antique plants already scattered through Europe and North America, or to investigate the existing garden documentation, including vital sources of information such as old plant catalogs, lists, pictures, and books. It is an exciting field, for now that so much of the world's native flora has been discovered, the past is the new frontier.

The Cottage Garden

It is an appealing idea: pushing open a wooden gate to find a front yard heady with perfume, buzzing with bees, with opulent rows of beans and marigolds, old cabbage roses, and caterpillar-free cabbages, and with a rickety seat under an ancient but heavily laden apple tree. Beside the cottage are planted beds of pinks or of old auriculas or "cottage" tulips, treasured for their smell or for the shape or markings of their petals.

Little is known about how medieval peasants in Italy, or Elizabethan ones in England, or even those of seventeenth-century France lived. Serious interest in how the owners or tenants of European cottages existed, let alone gardened, only really began in the early nineteenth century, and it wasn't until the 1850s, when some sophisticated and wealthy gardeners' taste for the very latest flowers began to get jaded, that a few started to rummage in cottage gardens for flowers that they remembered, or had read about, from the past.

Opposite: Elizabethan gardens were bright with all forms and colors of marigolds; here, they are mixed with nicotianas, rudbeckias, and others, and still make a pretty planting for cottage gardens. Above: Persian ranunculus are easy to grow and flower at the same time as the later tulips; a mix of the two can give a spectacular show.

Above: Hollyhocks, roses around the window, pinks, campanulas amongst the paving, and topiary birds make up this cottage-garden idyll, painted by Charles Edwin Flower in 1871 in the gardens of Kings Manor, Berkshire.

Though many cottagers turned out to be as interested in the new as anyone else, some cottage gardens proved still to contain splendid old plants. All sorts of interesting specimens turned up, from the ancient double yellow wallflower found by the Reverend Harpur Crewe (and which now bears his name), double forms of campion and campanula, old flaked carnations and pinks, curiosities like the "rose" plantain, ancient real roses, and antique daffodils to old fruit and vegetable varieties.

These discoveries fueled the cottage-garden ideal, one that is still powerfully alive. Not only are old flowers ardently collected and nurtured, but historic gardens are also being rebuilt and replanted in countries that have enough interest and resources. In general, though, "historic" also means "grand" and usually "well-documented." But small gardens must have been beautiful, too, though they can be glimpsed only occasionally in poetry, paintings, or ancient garden books.

The "cottage style" of gardening sometimes attempts historical authenticity, and indeed, as it is seen now — with a romantic "muzz" of plants, given form by paths; lavender or boxwood hedges; self-sown plants of columbine or Jacob's ladder; perhaps some topiary, certainly some old fruit trees — it may bear some resemblance to how old small gardens looked. British and American garden writers in the 1880s could have heard about the countryside gardens of their grandparents. Gardens can easily remain without rebuilding for almost a hundred years, so some late eighteenth-century cottage gardens were probably still in existence. But for anything earlier, gardeners have to rely on the books that began to appear at the end of the seventeenth century, written for the minor European gentry.

Important for the cottage-garden movement were works like the poetical *Paradisi in Sole Paradisus Terrestris*, written by John Parkinson in 1629; the rather more modest *The Scots Gardener*, written by John Reid in 1683; and the long sequence of editions of *The Gardener's Dictionary* by Philip Miller, starting in 1731. In all of them, and especially in Reid's book, the way of integrating formal layout with luscious planting, of mixing flowers with fruit and vegetables, is splendidly set out. From these books, too, come inspiring gardening ideas: smothering gateways with jasmine and honeysuckle, turning the corners in paths with the help of a topiary emphasis, and having pots of "greens" (a term that meant almost any tender exotic newly imported from Italy, China, or America) as the centerpieces to flower beds. Yet in many parts of the western world "cottagers" were so poor that thoughts of gardening anything more than vegetables for their own consumption would have seemed absurd. Perhaps, then, the modern image of cottage gardening is really a much faded picture of small manor-house gardens of long ago.

Whatever its origins, the cottage garden makes an exciting and productive way of combining a wide range of antique plants, satisfying around almost any sort of house, yet giving as much resonance of the past as any garden needs. Better, planting a cottage garden is fun. This is largely because there is no set way of doing it: it can vary from something kept reasonably trim, with topiary, brickwork paths, and antique flowers, to a floriferous tangle, unrestrained, perfumed and colorful, cheerfully combining cabbages, morning glories, jasmine, and dahlias.

Height can be given by fruit trees, and if you worry about thieves taking your crops, plant crab apples, myrobalan plums, medlars, or filberts — all delicious crops that will appeal to you, but probably not to them. Porches and doorways are always worth emphasizing: use some of the climbers detailed later in this chapter, or those suggested for the rose garden *(see p.102)*. For somewhere to escape from the kitchen or the phone, put up a rustic arbor, either as a feature in its own right or attached to the side of the garage, barn, or shed. Many companies sell ready-made metal and wirework ones, but something suitably rustic is easily made from scrap lumber, or wood and poles. In the nineteenth century, arbors were commonly covered with climbing beans or peas or climbing squashes and zucchini. They need to be deep enough to shade the seat on a hot afternoon.

Paths in cottage gardens give form to the planting and often look best when they go directly from one point to another. To give extra emphasis to the line, edge them with boxwood or clipped ivy. Brick makes a good surface, though flagstones or cobblestones look even better. Hard surfaces like these are more easily swept clean of cabbage stalks and rose petals than gravel. Lawns and grass paths are not part of a cottage garden, unless you are lucky enough to have an orchard.

The boundary fence or hedge is an important part of the design, keeping the world at bay and giving structure to the plot. Drape rails and fences with climbers or give the hedge a topiary flourish. If you live in the house year-round, give yourself a "florists corner" for some of the enchanting old flowers that look good in their own pots (old tulips, laced polyanthuses, or show auriculas). But best of all, make sure you have a sitting place for a summer evening, amid the heady perfumes of mignonette, stocks, nicotianas, jasmine, and honeysuckle.

Below: Brilliant anemones like this have been illuminating western gardens since the sixteenth century, and Middle Eastern ones for much longer. This one dates from about 1800.

Bulbs and Perennials

Among the bulbs and perennials of the cottage garden are some of the most cherished of all garden antiques. Some are to be found in every sales outlet; others have to be sought out. Some can be so vigorous that they will take over the yard; others need coddling if they are to flourish, but provide enormous rewards if they do.

Although a small cottage garden could easily be made using only plants from this chapter, you could try adding some roses, a boxwood topiary or two, and a fruit tree. Cottage gardens can easily provide interest for most of the year and be brimming with flowers from spring into late summer.

Acanthus ACANTHUS, BEAR'S BREECHES

Only two species of this handsome genus are antique garden plants. They have probably been popular since ancient Greece and certainly since Roman times. Both *Acanthus mollis* and the frillier *A. spinosus* were still widely grown in European gardens by the mid-1500s, though by 1754 they were thought worthy only of collectors' gardens.

In 1870, William Robinson, the English champion of wild flowers, extolled their virtues for the wild garden. Don't be put off; their extravagant foliage looks good in cottage ones, especially by a doorway or beneath an ancient apple tree.

Above: The leaves of *Acanthus mollis* have inspired designers from the architects of ancient Greece, where they were the source for the curling foliage at the top of Corinthian columns, to William Morris. The plant makes a dramatic focal point in the cottage plot.

Right: In a cottage border, dark maroon hollyhocks look good behind roses such as 'Ispahan' and 'Maiden's Blush'; fill in the planting with fennels, the taller bellflowers, and white tree lupines.

Alcea HOLLYHOCK

Hollyhocks have been in cultivation in Europe and Asia for so long that they have lost any connection with a known wild species. But the great age of the hollyhock really began in Britain and North America in about the 1840s. There were many famous breeders, though only an Englishman named Chater is remembered; his double-flowered varieties can still be found. In the cottage gardens of Europe and America, they were grown at the back of borders, often against walls, where the tall flower spikes were easily held in place and the basal leaves covered by other flowers. For an eighteenth-century look, grow the pink-flowered marshmallow (*Althaea officinalis*), once planted in cottage borders for late summer color.

Anemone ANEMONE

Of the 120 or so species, all from the northern hemisphere, the best have been in the garden for many centuries. For early spring, the 'St. Bavo' forms (*Anemone pavonina*) have been grown in Europe since the sixteenth century and were in Turkish gardens probably much earlier. Sir Thomas Hanmer, a famous seventeenth-century gardener, grew them mixed with the latest fraxinellas, new American tradescantias, and English primroses and daffodils. Copy that planting in your yard to give you flowers from early spring to midsummer.

Right: Introduced from the Far East, *Anemone hupenhensis* 'Splendens' was an essential feature of the mid- to late nineteenth-century cottage border.

Below: A semidouble white form of *Anemone nemorosa*, popular in the eighteenth century, is useful for brightening shady areas in a cottage garden.

Anemone coronaria is the 'de Caen' anemone; wild from the Mediterranean region to central Asia, it was first taken up in Western Europe by the florists of the French city after whom it is named. No early sorts survive, but double types now available make excellent substitutes for eighteenth-century gardens. They had been grown earlier, for in the early 1600s both types were crossed, with the handsome *Anemone* x *fulgens* the result. Of the richest vermilion and scarlet, it's still available.

For summer and fall beds, grow *Anemone hupehensis* (*A.* x 'Japonica' and other varieties) from China and Japan. All reached the West after 1844, though they had been oriental garden plants for centuries before that. Soft pink and perfect white forms were appearing in the 1860s.

Aquilegia COLUMBINE, AQUILEGIA

The fifty or so species come from all over the northern hemisphere. Before the mid-seventeenth century, only one (*Aquilegia vulgaris*) was grown in Europe. Variable in the wild, it flowers in violet, white, or red, and is sometimes bicolored. Single columbines can often be seen in medieval paintings and illuminations. They are still available.

One medieval painting shows a plant very similar to today's *Aquilegia vulgaris* 'Adelaide Addison,' so that form may be hundreds of years old. 'Starry columbines,' with flat star-shaped flowers, can still be found and may date from as long ago as the 1500s. *A. vulgaris* 'Nora Barlow' is of the same period. Many forms of this species have naturalized in North America, probably taken there by early colonists.

American gardeners of the seventeenth century were already fond of native species like *Aquilegia canadensis* and *A. caerulea.* John Tradescant had collected and was growing the glowing golden-scarlet *A. canadensis* by 1640. It was to remain popular; Thomas Jefferson grew it at Monticello, where it can still be seen.

All species and varieties of columbine are excellent in cottage-garden beds, easily grown, full of color, and self-sowing and crossing with abandon; they're especially good among irises and oriental poppies, though you could add species asters to give some late summer color.

Below: Garden pinks prettily marked like this have been grown since the late seventeenth century, when they were set at the sides of borders in spacious gardens and mixed with the buds of damask roses to make posies.

Bellis DAISY

There are fifteen species growing in Europe and North Africa. Forms of the common daisy (*Bellis perennis*) are strewn over missals and books of hours of the medieval period. By the late 1500s, fully double flowers in red or white, or with the two colors mixed, were popular and can still be found. *B. perennis* 'Alba Plena' is one, its double daisies packed with petals in purest white.

Other antique varieties still around include: 'Alice' and 'Dresden China' (perhaps eighteenth century); 'Rob Roy' and 'Robert' (nineteenth); and 'Aucubifolia,' with leaves flecked with yellow (mid-1800s). Several sorts were grown in North America by 1866, when *The American Gardener's Assistant* lists various types, including 'Hen and Chickens,' as greenhouse flowers. Seed strains are Victorian or later, fine for the loudest parterres. In cottage gardens use them to edge paths and flower beds.

Above left: The double columbine 'Nora Barlow' — a modern name on an ancient plant — makes an excellent border flower, either grown from seed or bought as a mature plant. Keep it away from other columbines to make sure it seeds true.

Above right: Double daisies, flowering in earliest summer, give color right at the front of the bed; in cold regions, overwinter them in pots indoors, where they will flower in early spring.

Dianthus CARNATION, PINK, SWEET WILLIAM, CHINA PINK

Carnations and the smaller clove gilliflowers (*Dianthus caryophyllus*) appear in medieval illuminations and portraits, where they are commonly double scarlet, with jagged petals and strong blue-gray leaves. Many flowers in sixteenth- and seventeenth-century illustrations resemble the gorgeous 'Fenbow Nutmeg Clove.' Once eighteenth-century florists got to work, endless new groups appeared, including "piquettes" (white, splashed with color), "flakes" (with irregular stripes of color radiating along the petals), and "bizarres" (streaked with three or four different colors). For a real antique, grow 'Raby Castle,' in salmon pink with darker flecks.

The basic species of pink (*Dianthus plumarius*) has small white or palest candy pink flowers, five fringed petals, and a delicious clovelike smell. Doubles of these types were popular medieval flowers, and 'Old Fringed White' and 'Old Fringed Pink' are still to be found. Try also 'Caesar's Mantle,' which may be from this period.

Feathered or "starre" pinks known in Britain in the seventeenth century had larger flowers, with frilled petals and often a darkish pink "eye." Many, like 'Ragamuffin' and 'Cedric's Oldest,' have survived in old gardens. A double form worth having is the deliciously scented 'Pheasant's Eye.'

In the eighteenth century, pheasant's eye types seem to have given rise to fully "laced" pinks, with red-purple edging to the white petal. The finest — 'Dad's Favourite,' 'Paisley Gem,' and 'William Brownhill' — may be early nineteenth century.

No *Dianthus* species likes being shaded by other plants, so keep them with border auriculas, low-growing thymes, and double daisies such as the pale pink 'Dresden China.' For vertical emphasis, add clumps of blue-gray-leafed bearded iris.

Sweet Williams (*Dianthus barbatus*) have been in cultivation in several forms at least since the early sixteenth century and are still vastly popular. A scarlet double was in cultivation by 1634; today's *D. barbatus* 'King Willie' may be a descendant. Single sweet Williams make good edging plants for a grass path through the vegetables.

Helleborus CHRISTMAS ROSE, HELLEBORE

There are twenty species of hellebore in cultivation in Europe and Asia, though some are rare. The oldest garden species are *Helleborus niger* (the Christmas rose) and *H. foetidus* (the stinking hellebore), native in all Western European woodlands. Though it may have had medicinal uses, the hellebore was a greatly admired garden plant in the seventeenth century but seems to have dropped out of use until the late nineteenth.

Helleborus niger, on the other hand, has been used by mankind since neolithic times, and probably as a garden plant since Roman ones. Prehistoric burials sometimes contain seed and capsules, but it may have been used as an arrow poison (it was still used as such by the Gauls) or for religious, medical, or magic purposes. The plant was a cure for madness, melancholy, and hypochondria well into the seventeenth century and was planted by cottage doors as a protection from spells into modern times. The seventeenth-century garden writer John Parkinson thought that the "flowers have the most beautiful aspect, and the time of his flowering most rare, that is, in the deeps of Winter about Christmas, when no other can be seen upon the ground." They all like shade, so plant them with other shade lovers such as sweet violets and pulmonarias and add con-trasting foliage from *Sanguinaria.*

Above: *Helleborus niger*, the Christmas rose, has probably been a garden plant since Roman times, but seeds and capsules have been found in neolithic burial places.

Opposite: The herbalist John Gerard grew the "orange tawney" daylily in the early seventeenth century. Now found as *Hemerocallis* 'Europa,' it is a vigorous plant, good by steps and doorways.

Hemerocallis DAYLILY

Of the twenty species of daylily from Asia, only two are antique garden plants. One is heavily perfumed and was once called "yellow tuberose"; the other is an ancient Chinese garden and culinary plant called "the flower of forgetfulness." The first is *Hemerocallis lilioasphodelus*, which reached Europe in Roman times and North America in the eighteenth century. Flowering early in the season and with narrow, rather floppy foliage, it makes a good plant beside garden steps or by a seating area. It is also grown in pots for the sunroom.

The other species is *Hemerocallis fulva.* The first form may have reached Europe in the 1400s and had brown-red flowers, though ones with red throats and one in pure yellow began to appear in the seventeenth century. The old form is now called 'Europa.' Doubles and bicolored flowers are all late nineteenth century, though 'Kwanso' types are old orientals. Several sorts were common in American gardens by 1890, and both here and in Europe daylily breeding began in the 1890s. Try pale yellow ones with bloodroots, ferns, and columbines, with peonies behind.

Above left: The sixteenth-century flag iris *Iris foetidissima*, with strange greenish-brown flowers, produces magnificent seedpods in late fall. A form with variegated leaves was popular in the eighteenth century and is well worth seeking out.

Above right: *Iris sibirica* relishes a damp place in the yard and has been a connoisseurs' plant for many centuries.

Iris IRIS

This is one of the great garden genera, with nearly two hundred species in Europe, Asia, and throughout North America. Several irises are among the most ancient garden plants. *Iris germanica* and *I.* 'Florentina,' the German or bearded irises, are probably the oldest. The first is in shades of blue; the second is silvery white and earlier flowering. As it flowers earlier than many irises, it combines with late tulips; I've often grown it among black parrot tulips, though a softer planting with yellow "cottage" tulips looks even better. Most other bearded irises date from after 1600, by which time gardens had a number of species. Nowadays, large numbers of bearded irises appear each year, making it a modern florists' group. However, irises have been bred since the 1880s, so antique ones can still be found.

Iris foetidissima has been grown in Europe since the sixteenth century, mostly for its splendid seed capsules, stuffed with orange-scarlet seeds. Try the variegated form in front of euphorbias, as foliage contrast in a drift of pale celandines.

Other suitable species for antique gardens include *Iris sibirica* (sixteenth century, though named forms are modern); *I. xiphioides*, the "English iris"; and *I. xiphium*, the "Spanish iris," all popular in the sixteenth and seventeenth centuries. The oriental garden species like *I. kaempferi* and *I. tectorum*, though ancient there, have been in western gardens only since the nineteenth century. The enchanting bulbous irises are also modern.

All the larger iris species are as valuable for their foliage as their flowers, making good structural accents amid the often hazy planting of cottage flowers. Try them with herbaceous peonies.

Lilium LILY

In a genus with seventy or so species growing throughout the northern hemisphere, two are among the oldest of domesticated plants, though they were first cultivated because their bulbs are edible (the bulbs are steamed before eating). They are the Madonna lily (*Lilium candidum*) from Anatolia, and *L. tigrinum* from the Far East.

In European gardens, the Madonna lily reigned alone until the early sixteenth century, though she did have several variant forms, including a now rare double. One or two American lilies attempted to usurp its place in the 1580s, and a few more in the eighteenth century, including *Lilium philadelphicum* and *L. superbum.*

Other essential cottage lilies include *Lilium martagon,* usually in mauve or purple, though the lovely white form was admired by 1634; and *L. chalcedonicum,* a sumptuous Turkish variety probably first grown by John Gerard from roots imported from Constantinople. Victorian flower arrangers combined it with the Madonna lily and black helleborine (*Veratrum nigrum*), and American gardeners had it by 1740. Try, too, *L. bulbiferum,* which has flowers ringed in orange and cerise but is tricky to grow. *L. pomponium,* a grand European, is also worth trying; John Rea boasted in the 1660s that his plants had eighty to a hundred flowers on each spike. However, nothing could be easier than a planting of ordinary martagon lilies among drifts of self-sown valerian, tangled with wild strawberries and ferns.

The first Asian lily (*L. dauricum* or *L. pensylvanicum*) arrived in the 1700s, a harbinger of a flood of Japanese, Chinese, and Korean lilies that appeared in the West in the nineteenth century.

Above left: This gorgeous spotted leopard lily (*Lilium pardalinum*) from California had been known in America for a century or more before it arrived in European gardens in 1875.

Above right: The Madonna lily (*Lilium candidum*) reached North America in 1630. Growing well in shade or sun, it is essential in every cottage garden. Try it with veratrums or ferns, good against a dark background such as an ivy-covered wall or tall hedge.

Paeonia PEONY

This is an Asian genus, with two American species and a few Europeans. Both the European species and several oriental ones have been grown since gardens began. Furthermore, gardeners on both continents have been envious of each other, so oriental peonies reached Europe in the early nineteenth century and caused a furor.

Paeonia officinalis, the "Female peionie," was single in ancient gardens; the lovely double "peony rose" was in every connoisseur's garden by the seventeenth century; and by the end of that century pink and white doubles, packed with petals and smelling just like the flowers of *Rosa gallica*, were available. Plant them in cottage borders, either by themselves or with honesty, pulmonarias, polyanthus, and lily-of-the-valley. All were cultivated in colonial America in the 1620s.

Chinese gardeners were compiling lists of tree peony, or moutan, varieties by 700 A.D. The first reached Europe in the 1790s. The plants are long-lived, and old varieties may yet be discovered. European collectors are once again searching Chinese gardens for ancient flowers.

The Chinese herbaceous peony (*Paeonia lactiflora*) was established in Europe only around 1800 but had many varieties in China by 1100 A.D., and one sixteenth-century nursery is known to have stocked thirty. In Europe they were crossed with *P. officinalis*, and by the 1850s there were endless gorgeous flowers. Surviving are 'Duchesse de Nemours' (1856) and 'Festiva Maxima' (1851), both double whites; 'Marie Crousse' (1892) and 'Baroness Schroeder' (1889), both blush pinks; and 'Madame Calot,' deep pink (1856). All flourish in shade; I grow them beneath gnarled apple trees, with veratrums, blue columbines, and a carpet of *Omphalodes verna*.

Papaver POPPY

This is a genus of about 100 species found throughout the northern hemisphere, from ephemeral annuals to the flamboyant huge perennial oriental poppy (*Papaver orientale*).

The Iceland poppy (*Papaver nudicaule*) was introduced in the early 1700s and *P. alpinum* — a relative of the more recent *P. burseri* — a few

decades later. However, *P. orientale* seems to have been in some European gardens by the 1600s and a common garden plant by the 1700s. Grow the single scarlet one among campanulas and artemisias, and all the forms of opium poppy (*P. somniferum*) you can find. The basic species has single flowers in white or mauve, but some of the brightly colored sorts may be 3,000 years old. The double white was grown in New England in the seventeenth century, and many garden varieties are mentioned in the 1700s, among them a double black, usually called 'Storm Cloud' (still obtainable), and a "double jagged Poppy, with beautiful striped flowers" (probably vanished).

If you want to use poppies in the yard, try a planting from Moorish gardens of the 1300s, especially admired for viewing by moonlight: grow them beneath medlars, pears, plums, cherries, and peaches, all twined with jasmine, and among the fresh green foliage of lupines, the lush glossy leaves of mandrake, white lily, and mint.

The oriental poppy does not seem to have been described as a garden plant much before 1714. The brilliant scarlet and black-purple centered one was the only variety cultivated until the early 1800s, when several related species arrived, including *Papaver bracteatum*, from the Caucasus and Persia in 1817; *P. pilosum*, from Asia Minor in 1852; *P. rupifragum* from Spain; and *P. atlanticum* from Morocco in 1889. These were crossed to give some of the exciting varieties listed today. Popular sorts from the 1870s include 'Nana Flore Pleno.' Many others, like the white and purple 'Perry's White,' are from the early 1900s.

A famous flower arranger of the 1860s, Miss Frances J. Hope, used crimson oriental poppies with straw and dark blue irises and mixed them with Aaron's rod (*Verbascum thapsus*), a combination that would make a fine antique flower bed.

Opposite: The double form of *Paeonia officinalis* (above) has been a garden plant since ancient times; the hybrid 'Sarah Bernhardt' (below) is from 1906 but is related to ancient garden flowers.

Center: Shirley poppies, bred around 1880, are selections from exotic forms of the wild field poppy (*Papaver rhoeas*), which has been in gardens since at least 1629.

Above: The opium poppy (*Papaver somniferum*), originally cultivated for its deliciously flavored seed and for the latex obtained from the immature seed capsules, also yields the drug opium.

Primula PRIMROSE, COWSLIP, OXLIP, AURICULA

This is another group that occurs throughout the northern hemisphere, with around 250 species, some much admired in the cottage garden. The primrose, one of the first flowers of spring, is painted in medieval manuscripts, and by the late 1500s dozens of variants were grown. The lovely, and now rare, double was the most valued by 1500. The double white, once again easily available, is also of the early sixteenth century.

The seventeenth century saw ever-stranger variants: hose-in-hose (with one flower inside another) and jack-in-the-green (whose sepals are enlarged to make a pretty green ruff). A number of similar types are still available.

Doubles remained popular; *Primula vulgaris* 'Lilacina Plena' (or *P.* 'Quaker's Bonnet'), easy to find and grow, is from the early 1700s, and the slightly darker 'Marie Crousse,' pink-violet edged with white, is from the early 1800s; it received an Award of Merit from the Royal Horticultural Society in 1882. All like shade later in summer, so they are good along a path through a berry patch.

The cowslip (*Primula veris*) has a garden history as old as the primrose. There were doubles by the time of John Gerard, whose famous *Herball* was published in 1597, and they continue to be mentioned throughout the eighteenth and nineteenth centuries. Some are still around, if rare.

The polyanthus is a complicated hybrid of *Primula vulgaris* (the word "polyanthus" was first used in 1683). A specimen was sent from Oxford, England, to Leiden, Holland, in that year, and in Europe it was called the English primula. Exciting variations soon appeared, and even in the seventeenth century, the beginnings of the "gold-laced" polyanthus were found. They were deepest red, with a narrow yellow band along each petal. By 1780, John Abercromby, a Scottish gardener, wrote that they were "one of the most noted prize flowers among florists." Seed arrived in the U.S. soon after, and the height of their popularity was reached around 1840, after which came a rapid decline. Good seed strains can readily be found. Try them beneath roses or by the path, where you can easily examine (or pick) the flowers. They look good planted among clumps of daylilies.

Garden auriculas are offspring of *Primula auricula* and *P. hirsuta.* By the end of the sixteenth century, gardeners had hybrids in white, red, yellow, and purple. Soon, there were those with a thick leaf surface of white wax (which gardeners called "meal"), some with a similar white "paste" in the centers of the flowers, and one with the beginnings of striping. Most seventeenth-century auriculas were grown in the open ground and were similar to modern border auriculas; the fancier varieties were grown in pots even then. In the eighteenth century, the pots were displayed on "stages" (rows of shelves covered with an awning over the top to protect the precious "meal" on leaves and flowers). By then, too, the first illustrations of the fascinating green-edged auriculas had begun to appear.

Border auriculas are difficult to grow in hot summer climates. In addition, show auriculas have flowers with "meal" on them, and this is easily damaged by wind and rain, so they need to be in a cold house or a special shelter. They grow well in any loam-based potting mixture. Watch carefully for root pests.

Border auriculas, with their subtle colors, shouldn't have to compete with other spring flowers. Give them space to themselves, then plant annuals such as yellow-green mignonette and blue bachelor's buttons among them. The auriculas will appreciate the summer shading.

Opposite: The modern green, gray, and white-edged auricula cultivars look like the antiques. They are still plants for show, not the open border.

Above: A typical color of a seed-raised border auricula, velvety and deep.

Below: A gold-laced polyanthus of a form that began to appear in the mid-seventeenth century. In good show forms, the lacing in the middle of each petal should reach to the flower's center.

Ranunculus BUTTERCUP, RANUNCULUS, FAIR MAIDS OF KENT

The genus has about 300 species, originating from Europe, the Near East, and North America. The showiest antique ranunculus, popular from the seventeenth century to the present day, is the half-hardy *Ranunculus asiaticus*. The first plants arrived in Europe in the late sixteenth century, some of them certainly from Syria. For modern antique gardens, most bulb suppliers list mixed lots of the 'Turban' sorts popular in seventeenth-century gardens. They're extremely fine, and are in flower with late varieties of tulip.

Among the hardy species, the most attractive, dating from the 1500s, is the double form of the white-flowered *Ranunculus aconitifolius*, called 'Fair Maids of Kent,' or 'Fair Maids of France.' This was common in eighteenth-century American gardens. Try it beneath *Syringa* x *persica* and with camassias and some of the species gladioli.

Above: *Ranunculus asiaticus*, an ancient inhabitant of Middle Eastern gardens has become a showy feature of western ones in early summer.

Tulipa TULIP

This is one of the great garden genera, with around fifty species originating from the Middle East. While some were known in Italy by the twelfth century, the first real garden tulips, produced by crossing between several species, were grown in Turkey in the early 1500s.

The first tulip to reach western Europe flowered in Augsburg in 1559 and was illustrated the next year. Even though it was a plain color, it caused a storm of interest. The striped sorts — whose coloration was due to a virus — soon followed, causing even more of a furor. The mania for grand variants was so strong that, in Holland in the next century, a "futures" market sprang up, with bidders paying colossal sums for bulbs they'd never seen and would sell on before they had a chance to grow them. In the 1630s, one bulb of 'Semper Augustus' (a variety which lasted into the eighteenth century) was sold for nearly 5,000 florins plus a new carriage and horses.

Valuable tulips were often grown in beds protected by plumed canopies, where their flowers lasted two weeks longer. Societies were formed for the breeding and showing of new sorts, though the vogue moved down the social scale, until tulip growing became the preserve of the artisan florists; the Wakefield Tulip Society, described as "old" in a British magazine of 1840, still holds a show every year, with beautifully striped tulips displayed in

beer bottles. The varieties shown there, being diseased, are not available in commerce. In fact, selling some varieties is illegal since the virus can infect not only tulips but other plants as well.

Non-virused sorts were also popular; *Tulipa* 'Keizerskroon' (1750) makes a cheery patch in the cottage borders, and the gorgeous *T.* 'Couleur Cardinal' (1840) is marvelous when densely planted in large tubs; it offers one of the most sumptuous reds of any flower. The "cottage" tulips, so called because many were found in cottage gardens in the late nineteenth century, are probably much older. 'Mrs. Moon' and 'Mother's Day' are favorites of mine; they do well planted as clumps in the flower bed and do not need lifting every fall. Black parrots, a seventeenth-century type, look splendid among brilliant orange and scarlet ranunculus; or try the sixteenth-century 'Lady Tulip' (*T. clusiana*) among young fern fronds and sedges around a small formal pool. A "tall yellow" form was popular in nineteenth-century American cottage gardens, and something similar would be good with green-flowered hellebores and palest yellow and white wallflowers.

Above left: A handsome mix of parrot and modern 'Rembrandt' tulips, without the virus that made old "broken" kinds so lovely. Here, oriental poppies will soon take over the show.

Above right: The strange, narrow-petaled *Tulipa acuminata* is probably close to early Turkish types.

Annuals

Above: Snapdragons grow easily from seed, but take cuttings of the best in the fall to increase your stock.

Opposite above: 'Art Shades' pot marigolds, fine antique doubles.

Opposite below: *Cheiranthus* 'Harpur Crewe, is ideal for small gardens, growing no higher than 12in (30cm).

Most annuals and biennials are easy to use and are essential in the modern style of cottage gardening, where they are allowed to self-sow, with the hoe removing only the worst placed plants. Many cottager-dwellers, particularly in the nineteenth century, used them in a piece of eye-catching spring and summer bedding. This was often in the front yard, to give a flourish of color and show to their houses, with perhaps a diamond-shaped bed or an arrangement of four hearts in a circle cut into the grass or outlined in glazed tiles. However you want to grow them, the group has lots of delectable antiques.

Antirrhinum SNAPDRAGON

The species most used in bedding is *Antirrhinum majus*, a European native that seems not to have been planted in medieval gardens but was certainly widespread in sixteenth-century ones; perhaps it had no medicinal use to interest early writers. Wild populations are very variable, and new forms reached the garden slowly over the next two centuries; various of these have been lost, including some with prettily striped flowers. In the eighteenth century, there were many named forms, propagated annually by cuttings. Although many were also grown in North America, all copy Thomas Jefferson, who had red singles growing at Monticello. For antique gardens, grow tall sorts from seed, in white, rose, red, or yellow; white ones, with blue bachelor's buttons, can look splendid between bushes of the rose 'Maiden's Blush.' Well-grown plants look good in pots, too.

Calendula officinalis POT MARIGOLD

Calendula is a quintessential cottage flower and so ancient that it has lost any contact with a wild species. It was common in medieval gardens all over Europe, and also known as 'Mary Gowle's,' 'Goldes,' or 'Ruddes.' At that time pot marigolds were used in the kitchen (to color soup and butter and to dilute saffron) and in the medicinal garden (its cordial cured depression). The plants were important enough to arrive in America with the first European settlers.

Doubles were popular by the late 1500s and come now from every seed pack. The strange little hen-and-chicken sort, with flowerets hanging off the main flower, is also still seen. Plants of the strain known as 'Art Shades' look good among parsley and mint, though in the flower bed try them with garnet-red violas and a feathery annual grass such as *Stipa pennata*.

Cheiranthus WALLFLOWER

Though there are American species, the garden wallflower (*Cheiranthus cheiri*) is a plant of the Eastern Mediterranean. The Romans must have brought it to western Europe, for it was in France and England by 1000 A.D. No doubt because of sentiment (it has been symbolic of faithfulness in adversity since the fourteenth century) and its lovely smell, it reached North America with the early settlers, though young plants must have had winter protection in much of the country.

Singles in brown and yellow (and both) were popular in sixteenth-century gardens, as were white, scarlet, and black in eighteenth-century ones. All can still be found, with modern names: one, 'White Dame,' is a fine greenish white. Doubles were in connoisseurs' gardens by the 1580s and were widely used in nineteenth-century ones. Two good doubles, propagated by cuttings, are still available from nurseries: *Cheiranthus* 'Old Bloody Warrior,' rich brown, large flowered, and heavily perfumed, may be sixteenth century; and *C.* 'Harpur Crewe' is a nineteenth-century name for a double form that would have been familiar to all seventeenth-century gardeners. It is covered in early summer with strongly perfumed flowers. Both make small bushes and look perfect with some of the decorative sages, among irises and beneath rock roses. Nineteenth-century Americans apparently also grew them in solaria.

Above: *Lupinus* 'Noble Maiden' is a modern plant based on a number of species that began to arrive in gardens in the early nineteenth century.

Opposite: *Lathyrus grandiflorus* is a splendid scrambling and perennial sweet pea. It is rather invasive, but the flowers have a sumptuous smell: perfect over a fence by the garden gate.

Lathyrus odorata SWEET PEA

Among more than a hundred species, this is the most heavily perfumed. In 1699, a Sicilian monk sent seeds to a British gardener living in Enfield near London. Next season, the London garden was filled with the perfume of the reddish-blue and purple flowers, and a garden story began. Although something similar can be found in medieval paintings, it was only now that gardeners became interested. Seed was sold commercially by 1724, and soon new sorts began to appear, including the pink and white variety called 'Painted Lady,' still essential in the cottage garden.

Several other old types are still around. 'Lord Nelson' is early nineteenth century and close to the original species. All are vigorous and perfect for growing in traditional ways — twining up obelisks or trellis or around the kitchen door. Grown thus, they were basic ingredients in American cottage gardens. It is also possible to plant seedlings in the open bed and let the plants scramble over roses (try 'Lord Nelson' through *Rosa* 'Madame Hardy') and up hollyhocks. If you can still reach them, deadhead until late summer, then let the pods ripen next season's seed.

Giant or 'Grandiflora' sweet peas date from 1877, but the modern ruffled 'Spencers' are all products of the twentieth century.

Lupinus LUPINE

Though there are three hundred species, almost all American, and though they are thought of as conventional inhabitants of cottage gardens, the large, colorful, and grandly scented lupines are all modern. However, some of the plain blue or purple seedlings are reasonably close to one of their parents, *Lupinus polyphyllus*, which arrived in Europe from North America in 1826. Breeding began in the 1880s, though the bicolored sorts date only from the 1920s. Antique gardens in Europe can grow the tree lupine (*L. arboreus*), introduced in the eighteenth century, which has yellow, sometimes white flowers; the more graceful *L. perenne*, taken to Britain by Tradescant the Younger from America in 1637; and the ancient annual *L. albus*, which was grown by the Romans for seed oil and admired for its scent. American gardeners do not seem to have grown any sorts until the late nineteenth century.

Tree lupines make floppy mounds of foliage rather than anything resembling a tree, so can also look good on the margins of a meadow. In the cottage garden, architectural plants make a good contrast — try the immense dark green leaves of acanthus or the elegant linear ones of irises.

All lupines make excellent meadow plants, and, as members of the pea family, their roots will add nitrogen to your grassland.

Above left: Pink-flowered sorts of the tobacco plant (*Nicotiana tabacum*) were once popular as balcony plants in summer, though they look as good in big drifts in the cottage border.

Above right: Plain orange forms of *Tropaeolum majus* have been garden plants since the seventeenth century and for sheer garden impact have hardly been surpassed.

Opposite: The late nineteenth-century *Viola* 'Irish Molly' needs careful propagation by cuttings in early spring to keep it going, but the color makes it worth almost any trouble.

Nicotiana TOBACCO PLANT

This is a large genus from Central America, containing some good shrubby species as well as the familiar herbaceous and annual sorts. The tobacco plant was among the first exports to England from the West Indies in the fifteenth century, but, though handsome and by the late 1500s available in several varieties, it was not often used in gardens. Over the next two centuries, it gained popularity as a way of covering less powerful — and less pleasant — urban smells.

In the nineteenth century, *Nicotiana affinis* and *N. sylvestris* appeared from the Americas and were first grown in France. The former is best in its white form, looking delightful with Japanese anemones and tall blue salvias. The perfume can fill the garden in late summer evenings. The even more elegant *N. sylvestris* was often used massed in large tubs, though it mixes well in shaded cottage borders, perhaps with peonies and ferns beneath apple trees. The scent is also delicious.

Tropaeolum NASTURTIUM

Of the numerous South American species of *Tropaeolum* (both annuals and perennials), the oldest garden plant is *T. minus*, introduced to Spain in the early sixteenth century. It is still represented by the late eighteenth-century double, 'Hermine Grashoff.' The species that tumbles across the paths of the *grande allée* in Monet's garden at Giverny, France, is *T. majus*, of the late seventeenth century.

Tropaeolum peregrinum is mid-eighteenth century, and *T. speciosum*, the flame flower from Chile, arrived in Europe in 1846. If you have evergreen hedges, put *T. speciosum* at their base.

Viola VIOLET, VIOLA, PANSY

Though we think of this as a quintessential nineteenth-century "bedding" group, the sweet violet (*Viola odorata*) had been cherished for centuries. Many named sorts date only from the late 1800s, though the 'Czar' is from 1863; it has the largest flower of all violets, in deepest purple. The leaves, especially after frost, are almost as perfumed as the flowers, so plant this variety next to the path. Mix with woodland strawberries for an even more delightful scent.

The classic pansies, violas, and violettas are complicated hybrids that first appeared in London around 1820, though they were partly parented from new American introductions. They were crossed, in the 1860s, into a charming species called *Viola cornuta*. The results were called violas. Named sorts are late nineteenth century; try 'Maggie Mott,' 'Bullion,' 'Pickering Blue,' and 'Irish Molly.' 'Violettas' are more recent; antique gardens are best with *Viola cornuta* in white, blue, or amethyst. The blue variant makes a fine foil beneath some of the sharper pink roses or covering the woody rhizomes of bearded irises.

CLIMBERS

Below: *Clematis* 'Jackman's Purple' caused a sensation when it first appeared in the 1850s but was soon widely thought of as a cottage flower.

The cottage porch or a ramshackle arbor, twined with woodbine, roses, and strands of jasmine, was part of the idealized picture that the Victorians had of cottage living. Nor was this just sentimentality. Ever since the sixteenth century, cottage gardens have been places for romantic tangles of climbers, disguising unsightly structures, clambering over trellises, or growing up a framework to shade the sitting areas and provide a shady and fragrant green arbor. They add to the pleasures of even tiny spaces, without reducing the ground needed for cabbages and herbs. The only modern addition to the cottage "look" is the vogue for letting twiners climb into fruit trees, though no ancient cottager would have dreamed of doing this; the fruit crop was too important.

Clematis CLEMATIS, OLD MAN'S BEARD

Despite there being 250 species of clematis scattered over the northern hemisphere, the only antique among the big-flowered hybrids is 'Jackman's Purple,' a nineteenth-century cross

Above left: The flowers of *Ipomoea purpurea* last only for a morning, opening from furled buds, but they occur in such abundance that that hardly matters.

Above right: Ivies like these can bring drama and interest to some of the shadiest parts of the garden, as well as turning blank walls into evergreen hedges.

between various species. Some grand gardeners hated these innovations and went back to old garden sorts, like *Clematis flammula* (sprawly, scented, perfect through eglantine roses) and forms of *C. viticella*. Both are sixteenth century; the double *C. viticella* is probably eighteenth. *C. alpina* was an occasional garden plant from the 1790s; *C. montana* is more recent (1830s). All look fine around windows, and do well on shaded facades or up rustic supports in the flower borders. *C. alpina* flowers just before apple blossoms open, so try some up an old fruit tree.

Hedera helix ENGLISH IVY

Hedera helix was one of the great nineteenth-century plants, draped over every mantel and sofa arch and every brown photo of "the dear departed." It had also been widely grown in the ancient world, where it was sacred to Bacchus. The yellow-berried "poet's ivy" (*H. helix* 'Poetica') was grown in Roman gardens and was equally popular in the sixteenth century, when it was hung outside taverns as a sign. It can still be grown. In the eighteenth century there were silver forms and a gold variegated one, but nothing else.

The nineteenth century saw a huge number of new forms, 'Irish' being the most popular. By this time, ivy had been equated with rustic charm and coziness and commonly covered cottages and villa porches. Both Irish ivy and new variegated sorts were popular for Wardian cases (see page 149). Later in the century, it became associated with tombs and churchyards and, like death, went out of fashion. Try ivies as ground cover, around a shady seat, or twined through a fence by the gate.

Ipomoea MORNING GLORY

Though this is a large genus containing shrubs and herbaceous plants, it is the sumptuous blue flowers of the climbing annual *Ipomoea purpurea*, and the cerise or white flowers of its relatives, that have been thrilling European gardeners since the late sixteenth century, when seed was imported from America. Thought to be a species of convolvulus, it was still called Indian bindweed in the eighteenth century. Rampant, it needs warm gardens, or shelter in cool ones. It is also good in pots, twined up a cane tepee. The only perfumed species is the white, night-flowering *I. alba*, often listed as *Calonyction aculeatum*.

Opposite: *Lonicera caprifolium* has a delicious scent, but it is prone to aphids. It makes a perfect cover for an unsightly wall or shed, in sun or part shade; here it is combined with *Clematis montana*, introduced from the Himalaya in the 1830s.

Below: *Jasminum polyanthum*, an ancient garden plant, is easy to grow, pestproof, and has a breathtaking perfume.

Jasminum JASMINE, ZAMBAC

Several species of jasmine, including *Jasminum officinalis*, *J. sambac* (the Arabian jasmine), and *J. grandiflorum*, have made gardeners swoon for many centuries. The wonderful, but not entirely hardy, white Chinese *J. officinalis* was grown in Persian gardens by the third century, when the Parthian rulers had extensive contacts with China. The plant soon became the symbol of Vohuman, the Zoroastrian archangel of good spirits, divine wisdom, and woman's loveliness. Jasmine spread around the Mediterranean region following the expansion of the Arab empire and was widely planted in Moorish Spain. It can still be found as one of the plants used for Spanish gloriettas (arbors of rose, jasmine, or cypress, used for dining). In seventeenth-century India, many gardens were designed for the night alone and filled with jasmine and the creamy-white-flowered tuberose (*Polianthes tuberosa*).

The jasmine probably reached non-Moorish Europe during the Crusades and was certainly widely planted by the 1300s. By 1650 good combinations were made in the gardens at the Villa Borghese in Rome, filled with roses, jasmine, tulips, amaranths, carnations, hyacinths, and jonquils. Try it. Among the white peacocks of Isola Bella, in Lake Garda in northern Italy, the early eighteenth-century trelliswork was once covered in jasmine, pomegranates, and oranges. Even in chilly Scotland, a seventeenth-century gardener suggested that your "gates should be twined about with jasmines and honeysuckle." A Boston gardener, Thomas Hancock of Beacon Hill, imported a few plants to the U.S. from an English nursery in 1736.

The only time that jasmine has been out of favor was in the nineteenth century, when one critic wrote that "the perfume of some flowers may affect breathing or digestion, or may cause dizziness and fainting." Don't let that worry you. In cold gardens, try the tougher *Jasminum* x *stephanense*. *J. officinalis* should be in every sunroom, where it is one of the easiest plants to look after.

Lonicera HONEYSUCKLE, WOODBINE

Of the many species of this climber, the European native hedge honeysuckle or woodbine (*Lonicera periclymenum*) has been in gardens probably since ancient Greece and certainly since early medieval times. It exists in several strains, two of which date from at least the seventeenth century. 'Early Dutch' was around every doorway and in the "wilderness." 'Late Dutch' (or 'Late Red') often flowers well into the fall and grew in the same places. Both are still essential garden plants. The other European native, *Lonicera caprifolium* (the perfoliate honeysuckle), was widely grown for its scent in the sixteenth century.

A Traditional Cottage Plot

This American garden, right next to a handsome old clapboard house in Maine, is probably on the site of the garden original to the house. It shows perfectly all the main elements of a cottage garden: the unaffected mingling of old and new; the mix of flowers, fruit, and vegetables; and the mix among the flowers of perennials, such as irises and lilies, with annuals or biennials, such as nicotianas and clary — the assemblage put together in delectable abundance.

Opposite: With its formal planning and enclosure, clipped topiary, and softly colored, fragrant plants, the garden nicely complements the old house.

Above: The classic cottager's plant *Lilium regale* is a strong presence in the garden. This is the most common form, with pure white, heavily scented flowers and petals backed in brownish burgundy; there is also an entirely white form, *L. regale* 'Album.'

Right: Late afternoon shadows show off the pale tones of lilies, annual cosmos, and the lovely old biennial clary (*Salvia sclarea* var. *turkestanica*). The planting of red astilbes and the climber *Phygelius aequalis* shows how simple colors and a mix of old and new flowers can make good and easy gardens.

Above right: *Clematis viticella*, or lady's bower, is a European native popular as a garden climber since at least the sixteenth century; gardeners could choose between several forms, including a double purple and a single red.

Left: *Nigella damascena*, the aptly named "love-in-a-mist," is allowed to self-seed around the yard. It can be sown in fall, or in early or late spring, for a succession of striking blooms throughout the year. The flowers are followed by attractively shaped seedpods.

Like all good cottage gardens, too, it is given a strong basic structure, first of all by its enclosure — here handsome, if modern, stone walls neatly palisaded along the top, a perfect place for draping the climbing plants — and second by its paths, which make a clear grid of linear brick walkways, as easy to walk on in a wet winter afternoon as on a dry summer morning.

In this garden, the enclosure and paths are combined in a rectangular plan that keeps the overflowing planting from turning into an out-of-focus mess. Fancifully curved paths may sound "cottagey," but they end up looking fussy. The strongly linear plan also means the garden yields visual pleasure even in winter, when all the plants except the topiary evergreens have vanished.

Above left: Remember to place your garden seating so that you benefit from the scent as well as the sight of your plants. In this garden, the dominant perfume is provided by the lilies.

Above right: Veronicas, daylilies, and *Eryngium bourgatii*. Veronicas like this were being introduced to gardens from North America and Asia in the eighteenth century; the eryngium was in European gardens by 1731.

The brick paths here have been laid with great ingenuity, the patterns varying in different parts of the garden to point out the different functions; the kitchen garden part, for instance, has the simplest pattern. The paths are edged with end-up bricks, a useful detail, as they give the margins of the paths more stability when the beds are dug over. Because the winters are severe, the owners have used engineering-quality bricks, which don't break up when the moisture in them freezes. However, cottage gardens can be given many other path surfaces: coarse sand or fine gravel, flagstones, or a pattern of different materials.

If you do use bricks, and you want them in a herringbone pattern as in parts of this garden, there is no need to use expensively cut bricks to straighten the edges: plant lines of thymes, double daisies, or border auriculas to disguise the ragged edge. With gravel, boxwood edgings are good, keeping the gravel firmly in its place.

The "kitchen" section of this garden has dwarf borders made of low espaliers of apples, great fun to look at though tricky to maintain. The main planting of fruit here is in a true orchard, but if your yard isn't large enough for that, then plant a fruit tree or two among the flowers, and as the tree grows, use more and more shade-tolerant flowers beneath.

The strongest scent in this highly perfumed garden comes from *Lilium regale*, not quite an antique (1903), but so easy to grow that it has almost eclipsed the other white lily, the much older *L. candidum*. For evening scent, the nicotianas, especially *N. sylvestris*, are unbeatable.

Colors are fairly muted, with a few splashes of brilliance, the most intense provided by the climber on the fence, *Clematis jackmannii* 'Superba,' developed by an enterprising plantsman, Mr. Jackman, in the 1850s.

The perennial flower planting provides good "bones" later in the season, with linear-leafed antiques such as *Iris sibirica*, some broader-leafed *I. germanica* hybrids, and daylilies, with the more softly structured nigellas, columbines, asters, centaureas, phlox, and herbs.

A cottage garden is also defined by its non-plant props: water barrels, birdhouses, pots, and seats. This garden is filled with charming ideas. Though you might not have a lead water cistern like the one pictured here, you can easily copy its enchanting planting in any big container: a mix of a dark purple-blue heliotrope (try 'Gatton Park') and the delightful peppermint-scented and velvety-leafed *Pelargonium tomentosum*. Both were favorites in Victorian times.

The garden also has a rustic seat — a style popular in the late 1800s and the early decades of this century — which looks especially comfortable among plants of the same date: *Salvia argentea*, daylilies, and the silvery seedheads of the Chinese *Clematis orientalis*, introduced to Europe much earlier, in 1731. However, almost any sort of seat, if it is made of a natural material, will be perfectly appropriate and give you somewhere to relax and enjoy the antique flowers.

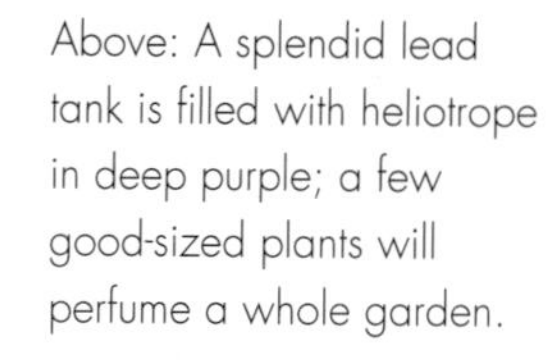

Above: A splendid lead tank is filled with heliotrope in deep purple; a few good-sized plants will perfume a whole garden.

Left: A low framework of espaliered apple trees makes a charming foil to the paths. The bricks are laid in patterns that differ throughout the various sections of the garden. In the foreground, bell jars make attractive cloches for young salad plants.

The Wildflower Garden

Throughout its history garden design has swung between the polarities of deliberate artifice and unembellished nature, or it has sought to accommodate both styles within the same plot. In ancient Rome, for example, gardens commonly had formal flower beds and topiary, but they also included areas that were supposed to be the mirrors of true wilderness.

In early medieval gardens, the "flowery mead" was a trim and sophisticated wilderness on a tiny scale, somewhere to sit in the warmth of summer, to flirt, eat, play music, or listen to poetry. During times of war and plague, it gave solace, as it did to the inhabitants of Boccaccio's *Decameron*, who fled to where "...in the middle of the garden was a square plot, after the semblance of a Meadow flourishing with high grass, herbs and plants, beside a thousand diversities of flowers, with orange and lemon trees, all around. At the centre a fountain of white carved marble had a gush of water, springing from a statue put on top of a column...."

Opposite: Meadow flowers, with their immense profusion and often small scale, make gardeners intensely aware of the subtlety of light, frequently in ways that the more brilliantly colored garden flowers do not allow. Above: Antique wildflowers of all sorts, from campions and soapworts to bluebells and lords and ladies, can show their full beauty in meadow and woodland gardens.

By the time of the Renaissance, the sophisticated flowery mead had gone out of fashion, but many grand gardens had wildernesses instead. The earliest example seems to have been built at Hesdin, in northern France, in the fourteenth century, and by 250 years later, every European palace and country house, from Versailles to Hampton Court and Chatsworth, had them, as did small manor houses like Tixall Hall in rural Derbyshire. The area near the house or palace was given over to topiary, knots, and parterres (whose intricate patterns could be appreciated from upstairs windows), and very visible artifice and cost; farther away, the wild garden or wilderness (decorated with "classical" elements such as fountains, pavilions, statuary, and arbors) was still very much a place to display wealth and learning — now often of decidedly pagan motifs and ideas — rather than somewhere to copy countryside and forest.

And so, with minor changes, it was to remain all over Europe until the early eighteenth century, when in England the pendulum of taste swung off course in a dramatic way. The landscape movement, with its very idealized vision of the wilderness, suddenly swept away obvious gardens, with their parterres, flowers, statues, and topiary. The model for the new style was not, as one might have expected, the wild countryside or farmland of Britain, but the landscape painting of a handful of artists who had lived mainly in Italy during the seventeenth century. It was all very grand, with scythed lawns sweeping up to the walls of the house and the lake kept neat and glittering. To add to the drama of the view — and to show off — grand temples, often modeled on fragments of Roman ones, and statuary were all carefully placed for the most painterly effect.

All this was fine if the garden owner had vast acreages of spare land but impossible to emulate for those with less. So, in about 1820, an intense reaction against "landscape" set in, and one strand of the reaction was a renewed passion for old-fashioned gardens, parterres, topiary, and all the artifice of the seventeenth century. It led eventually, amongst civilized if not rich gardeners, to the cottage garden movement. Another strand was an attempt to imitate authentic and picturesque nature without the trim lawns and classical buildings of the landscape garden. Though called "the picturesque" at the time, amid much heated debate, this movement soon gave rise to the wild garden, which, in the hands of its most publicity-conscious proponent, the magazine-owning William Robinson, made use not only of European wildflowers but of the species of all other continents, too.

The wild garden embraced a rather precious feeling for the untamed nature of forest or half-wild countryside, as well as a nostalgia for the rich flora and wildlife of ancient meadows increasingly being plowed up by farmers using new agricultural techniques and chemicals.

While poets have appreciated wild woodlands since ancient times, conservative gardeners have taken rather longer to see their beauties. The "wildernesses" described above would, to most modern eyes, have seemed entirely artificial. Even the more modest versions had the paths edged with wild strawberries, clipped ivy, boxwood, or even flowers like double primulas, hepaticas, or polyanthus.

Above: Bluebells (*Hyacinthoides non-scriptus*) painted by Beatrice Parsons about 1900. Other antiques, such as dog's-tooth violets, lungworts, lily-of-the-valley, and epimediums, can give equally delectable and perfumed sheets of color.

Opposite: Canterbury bells (*Campanula medium*), here painted by Redouté, have been a gardener's favorite since medieval times.

A closer approximation to real woodland was, sometimes, the "American garden," a vogue that started in Europe in the closing decades of the eighteenth century, when American tree and shrub species began to be grown in large numbers. Many of these, often of great beauty, had been collected along rivers, the easiest routes for exploring the continent. All, therefore, liked the same environment — damp, peaty soils — and so the American gardens tried to copy these conditions. Some followed the old wilderness formula, except that the path, under the influence of the landscape movement, went in curves rather than straight lines. On the other hand, many were left more properly wild.

The idea lasted well into the nineteenth century, and the flora gradually expanded, until it included rhododendrons from the East and from America, camellias from Japan, and so on. The look remained artificial, and only by the close of the nineteenth century were gardeners happy to have a piece of natural woodland and to plant among the shadows the lovely woodland flora from all over the globe.

However, interest in the older forms of "wilderness" is reviving and can offer an exceptionally attractive, labor-saving, and interesting way of gardening. Indeed, the woodland garden, like the sacred groves of ancient Greece, can be among the most satisfying and mysterious types of garden.

THE WILDFLOWER GARDEN TODAY

Wildflower gardening has moved on since the late nineteenth century, becoming green in the political sense and even hard-line by insisting on the rigorous use of native species alone. While that idea has some appeal, it seems to deny the fact that plants have always moved to new territories, and that the astonishing speeding up of this process over the last two hundred years must simply be accepted. There is no reason why a "meadow" in Europe should not have flowers or grasses from the Andes or the Himalayas, while one in Arkansas has plants from Central Asia or Central Europe. Or, more importantly, why a new meadow or woodland should not be scattered with antique flowers, wild or sophisticated, and still cherished, from gardening's own past.

Right: The harebell (*Campanula rotundifolia*) makes a delightful and vigorous meadow plant. It will prefer to grow in the more frequently mown places.

Opposite: The cornflower (*Centaurea cyanus*) is a good meadow flower in both Europe and America, particularly in the first few seasons: before the grasses get too dense, you can have drifts of its flowers.

The only shadow over the meadow garden ideal is that meadows, or prairies, were grazed by stock, which is impossible in the ornamental meadow. In a way, the medieval concept of the flowery mead is more apposite. Even if it is mown rather than grazed, the meadow can be a delectable place, requiring almost nothing in the way of regular maintenance, yet still be filled with narcissus and scillas in earliest spring, scattered with intense cerise corn flags and milky blue camassias in summer, with the last of the butterflies flittering over meadow saffron at the end of the season.

Your woodland garden, wilderness, or fruitful orchard, can be carpeted with wild strawberries, sweet violets, and bloodroots and be filled, through almost the entire year, with the perfumes of daphnes, sweet box, musk roses, lilac, and honeysuckle, while its paths are edged with snowdrops and wild primulas. What a bounty.

If you're starting a meadow garden from scratch, spare a thought for the grasses that will, after all, be the main bulk of the planting. A large numbers of grass species will be native to the area where you live, and for the planting to look authentic, as many of them as possible need to be represented. Grass seed mixtures designed for lawns will hold almost none of them, so, if at all possible, sow a wild grass seed mixture harvested from local meadows. If you are starting from an already half-wild piece of grassland, establish the flowers as almost mature potted plants and remove the grass for a foot or so around each new plant. If you merely scatter wildflower seed over grassland, almost nothing will germinate.

MEADOW BULBS

Below: *Camassia leichlinii*, in shades of blue from deep and thrilling to pale and milky, soon makes strong clumps that can be divided every few seasons to stop them from getting cramped.

Many bulbs are natural denizens of meadow and woodland, spending inhospitable seasons below ground. These dormant periods are used to escape drought, or low light levels when the surrounding trees are in full leaf, or cold. Bulbs can enliven your greensward and your grove in earliest spring. Keep the former scattered with summer color until the first mowing and brighten fall in the latter as the branches above begin to scatter their leaves; to see a piece of light autumnal woodland filled with cyclamen and meadow saffron is one of the great excitements of gardening.

The plants don't have to be special: a woods filled with bluebells or the trumpet narcissus can be as breathtaking as if it were scattered with rarities. Leave mowing the meadow until most of the bulb foliage has yellowed — for narcissus, that's at least six weeks after flowering; for some alliums, it is only a week or two; and for things like camassias, it seems to depend on the season.

All the bulb species below are tolerant of rough-and-ready planting, and it is usually simplest to lift a divot with a spade, nestle a bulb or two into the soil beneath, then replace the sod. Large bulbs such as *Fritillaria imperialis* need to have a hole excavated for them so that the sod can be replaced. Firm the sod down but not hard enough to damage the bulb.

Though many of the bulbs described here will begin to colonize your meadow in a season or two, plant as generously as you can at first so that, in the area devoted to each species, you start off with a reasonable show. In all but the tiniest meadow, and with any but the largest bulbs, any planting of less than 50 per "patch" will hardly be visible. Keep the patches looking as natural as possible by spacing bulbs closely at the center, perhaps a foot apart, and scattering them more distantly at the margins.

Camassia CAMASSIA

The five species of camassias, all American in origin, grow easily in moist grassland and have spires of starry flowers, in shades of blue that can vary from deep sapphire to milky pale or, in some forms of *Camassia leichtlinii*, to palest cream. The bulbs of *C. quamash* were a food crop for Native Americans, though they must also have admired the dark blue flowers. In European gardens from

Above left: *Colchicum speciosum* 'Album,' one of the grandest whites, is a strong grower, with large summer leaves. Look for seedpods once the leaves have unfurled; seed takes two winters to germinate.

Above right: *Crocus tommasinianus*, one of the earliest crocuses to flower, makes a dazzling start to the season and will increase well.

1827, there are several garden forms. Taller and grander, *C. leichtlinii* arrived a few decades later. All its forms are worth having, and there's a very special double cream one. All but this variety produce seed; sow it in pots if you need more plants and keep seedlings potted for a season or two before planting out in the meadow.

Colchicum MEADOW SAFFRON

There are sixty or so species of meadow saffron, all Old World; they are also called autumn crocus but are not closely related to fall-flowering species of the genus *Crocus*. They have fascinated European gardeners since Roman times. The one widely grown then, and all over medieval Europe, was the native *Colchicum autumnale*. Singles and doubles in rosy pink, purple, and white were once grown and can still be found. At least one form was taken to North America by the first Dutch settlers. All look lovely studding the greensward that grows back in after the last mow of summer.

Grander varieties arrived in the sixteenth century from Constantinople, where they had been popular for many centuries before that. *Colchicum byzantinum* is small-flowered but worth looking at; plant it by a path so that you can see its checked petals. For meadows in the nineteenth-century style, grow varieties of *C. speciosum*: all have large flowers, and the white one is a marvel. Don't mow until the leaves shrivel away in early summer. Divide the clumps every few seasons or they will stop flowering.

Crocus CROCUS

Saffron, the autumn-flowering *Crocus sativus*, has been grown since the beginnings of civilization and has lost touch with any wild species. It will naturalize in warm and sunny grassland but is not worth trying in gardens with short summers. Elsewhere, for spring, try the lovely, honey-scented *C. susianus* (syn. *C. angustifolius*, cloth-of-gold crocus), popular since the sixteenth century, or the enchanting *C. imperati*, in favor since the seventeenth. Most other spring crocuses are modern hybrids, though *C. tommasinianus* is just antique and naturalizes easily. For the fall, *C. speciosus* is equally easy and occurs in several varieties, notably 'Artabir' with deep violet-mauve flowers.

Fritillaria FRITILLARY

Though there are some beautiful American species that would do well in a wildflower garden, the most ancient garden fritillarias are European or Near Eastern: *Fritillaria meleagris* (the snakeshead fritillary) and *F. imperialis* (the crown imperial). The snakeshead, with its petals checked in purple and cream, is an easy meadow plant. Flowering early, it can make a fine, if subtle, show. If you want more impact, plant the white variant, which is also sixteenth century.

The impressive *Fritallaria imperialis* has been grown since the late 1500s. It does best in shade and in warm gardens. Though handsome in flower, it has an unpleasant scent, so do not plant it near the house. The lemon-yellow *F. imperialis* 'Lutea' looks best in meadow grass. Some varieties reached America in the late 1600s and may have been grown by Dutch settlers even earlier.

The black-flowered *Fritillaria persica* from Turkey and green *F. pyrenaica* have been gardened since the seventeenth century and will do well in meadows in warm climates. All other species are recent arrivals in the garden.

Narcissus DAFFODIL, NARCISSUS, JONQUIL

This is an extraordinary genus, an assemblage of only thirty-five species, but thousands of hybrids. It is filled with antiques, from the "sacred Chinese lily," one of the tazetta types that reached Europe from ancient Egypt or Greece via China, to the 'Van Sion' daffodil grown by the Tradescants. The easily naturalized trumpet narcissus (*Narcissus pseudonarcissus*), with its pale yellow twisted perianth lobes and deeper yellow trumpet, has been grown since the sixteenth century.

All the "poet's narcissus" or "pheasant's eyes" are another group from the ancient world; they do well in the wild garden and smell delicious. The largest meadows should have only a handful of varieties, preferably spread through the flowering season. An egg-yellow sort such as 'Van Sion' looks good with pale cowslips, even better with oxlips, and a scattering of fritillaries. The late-flowering poet's narcissus is a perfect partner to the first campions.

Above: Plant the poet's narcissus by a seat or path and enjoy the scent.

Below: *Scilla peruviana* brings color to summer meadows, as it did in the eighteenth century.

Above and left: *Fritillaria meleagris* was used by Elizabethan ladies to decorate their dresses; their Dutch counterparts in New Holland may have done the same. It went on to become a popular cottage-garden flower.

Scilla SQUILL

Though antique gardens once sported the pretty flowers of *Scilla bifolia* and *S. peruviana*, the antique most easily established in the meadow garden is *S. sibirica*, which has been popular in European gardens since the eighteenth century. Once established, it seeds very readily, and few gardeners can ever have enough of its glorious deep blue flowers at the high point of spring. A white form of *S. peruviana* was much prized in the eighteenth century.

The European bluebell is now placed in another genus (*Hyacinthoides*); it makes an easy and sometimes spectacular meadow plant.

MEADOW ANNUALS AND BIENNIALS

Surprisingly few annuals grow in meadows, where the dense growth of grass makes their lives hard. Poppies, feverfew, and fumarias are really weeds of disturbed or plowed soil. Still, try a few. They're best established as part-grown potted plants, then set out in a patch of bare earth. Biennials, too, can make a good show, starting with the purple money plant (keep them well away from the brighter yellow sorts of daffodil or plant varieties with white flowers) and carrying on into the summer with perfumed drifts of *Hesperis matronalis*. All of them are best planted near tree trunks or the shrubby margins of your meadow.

Hesperis SWEET ROCKET, DAME'S VIOLET

The billowing sprays of white or mauve scented flowers of the biennial *Hesperis matronalis* were much admired in medieval gardens; a double white form appeared in the sixteenth century. This is still around, but tricky to grow and keep over winter; it is something to be cherished in the flower garden, not planted in the meadow. The singles seed readily, so leave the flowering shoots to mature into late summer.

Lunaria MONEY PLANT

Honesty (otherwise known as silver dollar, white sattin flower, balbonac, penny flower, silver plate, honesty, pricke-sangwoort, sattin) was a favorite of medieval gardeners and was even mentioned by Chaucer, who is not usually regarded as a source of garden information.

Now *Lunaria annua* is commonly found in hedges and meadow gardens. There are a large number of variants within the species: the flowers range from burgundy to white, with lots of reddish-purples in between, and the leaves are sometimes dramatically variegated, making the plants look almost too busy. The fleshy roots of *L. annua* were once eaten in salads, perhaps explaining why money plant was one of the first European garden plants to reach North America.

Lunaria rediviva, a perennial, had the same uses and does well in the wild garden, though it prefers more shade. It's a good plant for the difficult area where meadow blends into woodland. The pale violet flowers have a scent that carries well on the spring air. The seedpods, pointed and elliptical, are less suited for drying.

Oenothera EVENING PRIMROSE

Most species of *Oenothera* are American, though there is a Tasmanian representative. They have become naturalized in every other continent and can be troublesome weeds. Don't risk them in the flower garden; let them take their chance in the meadow, where their large yellow flowers, often nicely scented, look good. *O. stricta* (sometimes called *O. odorata*) arrived in 1790, and the short and perennial *O. acaulis* was used as a decorative edging plant in Victorian shrubberies and wildernesses. It can be established in the meadow.

Above: Purple- or white-flowered *Hesperis matronalis* provides one of the essential scents of a summer afternoon; plant it anywhere you are likely to put a seat.

Left: Once seeds and outer casing have been shed, the translucent membranes of the money plant's seedpods make good decoration well into winter and last even longer indoors.

Opposite: Evening primroses are too invasive for cottage borders, but make perfect plants for wild areas and meadows. This is *Oenothera biennis,* which was sent to Padua from Virginia in 1619.

Meadow Perennials

Though lady's smock and cowslips will be in flower with the last of the bulbs, most meadow perennials flower in summer. All are best planted in discrete drifts, with initial plants at 3ft (1m) intervals and with a few others more distant. Plant the smaller ones close to mown paths, allowing larger plants to register from farther away.

If you are planting directly into established grassland, watch all new plants carefully throughout their first season. Grasses are competitive and will steal water and light away if they can.

While some tall perennials will tolerate mowing in late summer (plants like *Lythrum salicaria*, *Solidago*, *Echinacea*, and *Papaver orientale*), the best ones are low-growing or have flat rosettes of leaves that remain beneath the mower blades.

Cardamine LADY'S SMOCK

Cardamine pratensis is essential in the meadow, where its four petals shine amid the young grass of late spring. Though it seems to do perfectly well in ordinary grassland, if you have a dampish meadow, try the much grander double forms. There are several minor variants, in slightly different shades of mauve, enchanting and popular since the mid-1700s. If you want something white, try *C. trifolia*; gardened since the sixteenth century, it will give you drifts of tiny white flowers where meadow joins shrubbery or wilderness.

Primula PRIMROSE, COWSLIP, OXLIP

Though these also appear in the cottage garden (see page 33), the simpler sorts of primrose (*Primula veris*), cowslip (*P. vulgaris*), and the greenish-yellow oxlip (possibly a hybrid between them, but given the name of *P. elatior*) deserve a place here, too. Try the rosy-purple form *P. vulgaris* ssp. *sibthorpii*, especially by a mown path or rustic chair. It is extremely floriferous, an unusual color so early in the season, and has been grown since the sixteenth century.

Silene CATCHFLY, CAMPION

Many of the catchflies and campions make good meadow plants. Ragged Robin (*Lychnis flos-cuculi*) is especially good, and you can find an amusing white-flowered variant. For perfume, try the closely related soapworts (*Saponaria officinalis*), in marshmallow pink or white, single or double. Mown after flowering in midsummer, the plants soon make splendid and strong-growing clumps.

Opposite: Red campions make easy meadow plants, grow well on shady banks, and look lovely in long grass beneath old apple trees.

Below: Cowslips (*Primula vulgaris*) establish well in dampish meadows; in this one, it is grown with the pale lilac *Cardamine pratensis* in its wild single form. The doubles are pretty, too, but do best in the cottage garden.

Woodland Flowers

So much of the garden is designed for sunshine and summer that it is easy to forget that tree species are the natural dominant plant type over much of the land area of our planet. Huge numbers of smaller plants have evolved growing among or beneath them and have of necessity learned to cope with varying degrees of shade, intense root competition, even drought. In temperate forests, where the trees are mostly deciduous, the subsidiary plants often produce their own foliage and flowers early in the year, before the branches above become thick with leaves.

Among this vast woodland flora are to be found some of the most delectable of all garden flowers, and some of the finest are true antiques in the garden. All gardeners should try to grow at least some of those listed below, and those with shady gardens — whether woodland or urban shade — should plant everything they can find. The rewards, for surprisingly little effort, are enormous.

Aconitum MONKSHOOD, WOLFBANE

The aconitums of evil reputation (their juices were once used to tip poisoned arrows and for baiting a lamb's carcass to destroy wolves) make splendid flowers for light woodland, especially if the soil is damp. The first into cultivation seems to have been *Aconitum napellus*, whose flowers were depicted in medieval manuscripts and which became a popular garden plant by the Renaissance. Seventeenth-century forms had pink (*A. napellus* 'Carneum') or white (*A. napellus* 'Albidum') flowers. *A. septentrionale*, the yellow wolfsbane, is also medieval, but Victorian wild gardens often combined it with "gardeners' garters" (the striped-leafed grass *Phalaris arundinacea*) and the droopy white plumes of *Aruncus sylvester*. North America has its own species of aconite (*A. uncinatum*), introduced into Europe in 1770, though the European ones were certainly in American gardens by the nineteenth century.

Right: *Aconitum napellus* should be planted in large clumps for maximum impact and the roots divided every few seasons.

Far right: None of the varieties of *Convallaria majalis* can rival the single white form, with its glorious perfume.

Above: Cyclamens seed themselves easily once a few plants are established in woodland without much undergrowth. Sheets of them can make astonishing sights in early fall or spring.

Convallaria LILY-OF-THE-VALLEY

Lily-of-the-valley (*Convallaria majalis*), with its tiny white or pink flowers, is an ancient garden plant, admired since at least 1000 B.C. Unsurprisingly, there are many variants. Pink- and red-flowered forms were popular during the Renaissance, an inelegant (and still available) double-flowered white was grown by 1770, and ones with leaves variegated in white or yellow appeared a century later (variegations fade as summer progresses). The old white form was popular in eighteenth-century North America, but seems to have gone out of fashion in the 1880s. Though they look good by woodland paths, plant a few wherever you have a sitting area, in sun or shade. Then savor their scent in early summer.

Cyclamen CYCLAMEN, SOWBREAD

The old "sowbread" is *Cyclamen coum*, grown first probably in Roman times as a medicine for pigs, but then as a garden plant since the sixteenth century. Other hardy sorts have been connoisseurs' plants since the seventeenth century, all popular because they flourish in the driest shade beneath shrubs and trees where little else will grow. They make a memorable sight in sufficient quantity, rendering the shade rose purple in either fall or spring. They seed themselves readily once several corms are established.

All were grown in America by 1725 and are popular woodland garden plants today. In my garden in southern Scotland we grow them with polypody ferns and epimediums.

Epimedium EPIMEDIUM

Elegant ground cover beneath trees, many epimediums are antiques. The oldest, *Epimedium alpinum*, was in Gerard's London garden in 1597 (he acquired it from the French king's gardener). The others are all nineteenth century. *E. grandiflorum* reached Europe from Japan in 1830 and *E. pinnatum* arrived from Persia in 1849. Leaves color prettily in the fall in shades of pink, coral red, and bronze and look good among ferns, daylilies, and willow-leafed gentians. Clear foliage away in spring to get a good view of the charming airy sprays of yellow or rosy orange flowers.

Eranthis WINTER ACONITE

No woodland spring is complete without the lovely *Eranthis hyemalis*, easily established from bought tubers and soon seeding itself into the surrounding area. Re-establish the seedlings into masses of aconite to suit whatever space you have available. The foliage dies away by summer, allowing room and light for the other woodlanders that emerge later in the season.

Aconites were plentiful in London gardens by 1597, though they may have been recent arrivals. They look particularly good planted among the early-flowering *Helleborus corsicus* and *H. foetidus*.

Above: The winter foliage of epimediums makes a tough and subtly colored cover in light woodland. The plants spread vigorously in good soil.

Right: The flowers of the dog's-tooth violet are so exquisite that you should put them where you can best appreciate their beauty. Seedpods mature quickly after flowering, so keep watch if you want to collect seed.

Erythronium DOG'S-TOOTH VIOLETS

Nothing to do with real violets, *Erythronium dens-canis*, from Eastern Europe and Asia, was in gardens by the sixteenth century and existed in many forms of leaf mottling and flower color by the eighteenth. The Tradescants grew the first American species, *E. americanum*; others, like *E. revolutum*, *E. tuolumnense*, and *E. oregonum*, had to wait until the 1860s. All make elegant and easy woodland plants, flowering in spring; the leaves vanish by the height of summer. It is hard to have enough of *E. dens-canis*, with its marbled foliage and pendant, cyclamen-like flowers; harvest and sow the seed as soon as the capsules ripen. Leave the pots of seed outdoors over winter. I grow them among bloodroot, wood anemones, soft shield ferns (*Polystichum setiferum*), and the pale yellow and perfumed daylily 'Hyperion.'

Above left: The brilliant greenish yellow bracts surrounding the tiny flowers of euphorbias make glowing additions to the woodland garden.

Above right: Once established, winter aconites naturalize easily, seeding themselves with vigor. Seed germinates as mature plants flower, so don't weed around them too thoroughly or you will clear away next year's crop.

Euphorbia EUPHORBIA

The English garden designer and writer Gertrude Jekyll was wildly enthusiastic about her *Euphorbia characias* ssp. *wulfenii*: "the immense yellow-green heads of bloom are at their best in May, they are still of pictorial value in June and July...." It is indeed lovely, likes shade, and will seed itself around gently.

Like several other popular species such as *Euphorbia robbiae* (with lime-green flowers) and *E. epithymoides*, also called *E. polychroma* (whose flowers are bright yellow), it was a nineteenth-century introduction. Several others are much older. The strange caper spurge (*E. lathyris*) is a garden weed worldwide. The lovely green-yellow bracts of *E. palustris* are to be preferred and are also medieval. Try this plant with pale yellow flag irises (in flower as the euphorbia's bracts darken), white narcissus, and a ground cover of ivy.

Above left: The elegantly pleated leaves of the various *Veratrum* species unfold from large green "snouts" in late spring and offer the most dramatic foliage of the season.

Above right: *Omphalodes cappadocica*, with larger leaves and flowers than *O. verna*, dates from 1814. Flowering from early spring, it looks pretty around a woodland pool.

Omphalodes BLUE-EYED MARY

The blue-flowered *Omphalodes verna* is native to shady woodland in southern Europe. It reached English gardens in the seventeenth century and was used for edging paths in the wilderness, where it stills look good (I grow it beneath sarcococcas and tree peonies). *O. verna* 'Alba' is an enticing white-flowered variant

Sanguinaria BLOODROOT

The gorgeous *Sanguinaria canadensis*, with its frilled and plate-sized green-gray foliage and immaculate white flowers, was grown in Europe by 1680, and the even lovelier double was around by a century later. Eighteenth-century gardeners grew it with dog's-tooth violets, spring cyclamen, and small irises. The combination still works. The sap resembles oozing blood and was used as body paint by Native Canadians. Jefferson grew the plant in quantity at Monticello.

Veratrum VERATRUM, FALSE HELLEBORE

The large and elegantly pleated foliage of *Veratrum nigrum* and *V. album* has enchanted gardeners since the sixteenth century at least, though eighteenth-century ones complained that they attracted slugs and snails. Still, watching the huge noses appear in spring and expand into great spear-shaped leaves makes them worth the trouble of pest control. Flowers are grayish white in *V. album* and deep purple in *V. nigrum.* All look good against contrasting foliage, so plant among ferny-leafed sweet cicely, euphorbias, and glossy-leafed *Daphne pontica*, with periwinkles below.

Vinca PERIWINKLE

The periwinkles, whose flowers were often scattered across the pages of medieval illuminated manuscripts, make splendid ground cover for woodland plantings, whether formal or informal.

In the woodland garden, it is best to use the small *Vinca minor*, ideally the double-flowered blue or purple, both at least seventeenth century and both giving dense cover. The larger *V. major*, though sprawling wildly, tends to more open growth and therefore allows weeds you might not want to force their way through.

Left: With flowers like tiny double water lilies, the double-flowered bloodroot (*Sanguinaria canadensis*) also has sumptuous bronze-green foliage once the flowers are over.

Below: Their deep glossy green foliage (variegated in some antique forms) makes vincas good for underplanting even in quite dense shade.

WOODLAND SHRUBS

Above: *Philadelphus coronarius* 'Variegatus,' a delightful bush of the eighteenth century, does best in light shade.

Below: The hydrangea-like head of the guelder rose, perfect in spring, goes on to produce translucent scarlet berries.

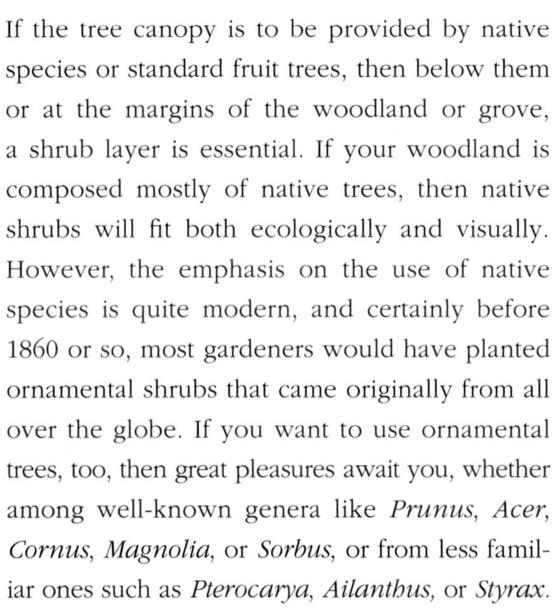

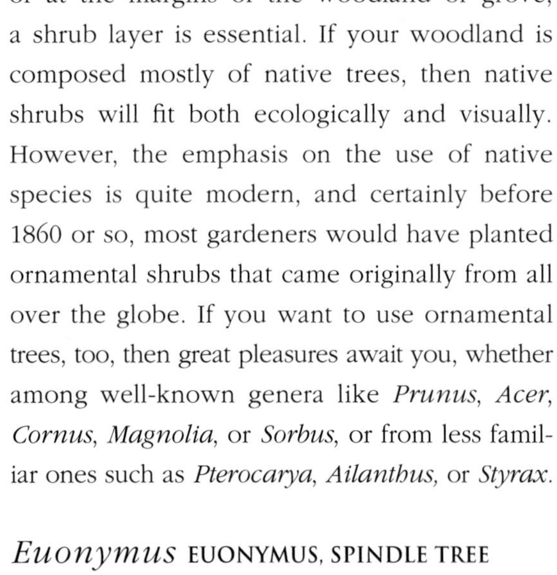

If the tree canopy is to be provided by native species or standard fruit trees, then below them or at the margins of the woodland or grove, a shrub layer is essential. If your woodland is composed mostly of native trees, then native shrubs will fit both ecologically and visually. However, the emphasis on the use of native species is quite modern, and certainly before 1860 or so, most gardeners would have planted ornamental shrubs that came originally from all over the globe. If you want to use ornamental trees, too, then great pleasures await you, whether among well-known genera like *Prunus*, *Acer*, *Cornus*, *Magnolia*, or *Sorbus*, or from less familiar ones such as *Pterocarya*, *Ailanthus*, or *Styrax*.

Euonymus EUONYMUS, SPINDLE TREE

Euonymus europaeus, the only European of this fascinating genus, most of whose species come from North America or Asia, was popular in the wilderness, pruned into a small tree. The red autumn leaves and curious scarlet capsules still make it an exciting plant to grow. By the eighteenth century, *E. americanus* (discovered in 1683) and *E. verrucosus* were becoming popular alternatives. All are unjustly neglected nowadays and should be used much more widely for small shaded yards, as well as in the woodland.

Philadelphus MOCK ORANGE, PHILADELPHUS

Many of the hybrid mock oranges were bred in the late nineteenth century, as Chinese and North American species began to arrive in Europe in quantity. Only the European *Philadelphus coronarius*, as splendidly perfumed as any of them, is an old garden plant. It is one of the great smells of summer, enjoys woodland conditions, and can be left to its own devices. If you want more excitement than the basic species offers, look for the splendid variegated form of the 1770s.

Skimmia SKIMMIA

Early nineteenth-century urban gardeners throughout industrialized Europe were always desperate for plants that would survive the intense air pollution, preferably ones that were evergreen and had brilliant berries. *Skimmia japonica*, introduced

in 1838, had all these qualities, together with powerfully perfumed flowers, and would grow in the densely shaded backyards of close-packed houses. It will, of course, do far better in the woodland garden. Put it beside a sitting area for the end of a winter afternoon stroll.

Viburnum VIBURNUM, GUELDER ROSE

The guelder rose (*Viburnum opulus*) or "pliant mealy tree" as it was called in the eighteenth century, flowers in winter; in fall it bears trusses of scarlet or yellow berries and the leaves color richly. No wonder it was in all old gardens from the sixteenth century and earlier and in wildernesses from the late seventeenth. It should be in all modern ones, too. *V. tinus*, the well-known evergreen laurustinus, shares the same history, though it was often grown in pots to perfume the hall or orangery in winter. Outdoors, in all but the coldest gardens, it makes dense bushes of glittering, dark green leaves, a perfect foil for the fall colors of maples and cherries.

Above left: Many of the euonymus color well in fall; this is *Euonymus europaeus*, popular in wilderness gardens for two centuries.

Above right: The brilliant berries of skimmia last through the winter, soon to be followed by a new crop of heavily scented flowers.

A Meadow Garden

Above: Meadow salsify (*Tragopogon pratensis*) flourishes in meadows. Large "dandelion" seedheads follow on from the flowers. The roots are edible.

Below: Oxeye daisy (*Leucanthemum vulgare*) gives easy drifts of pretty white flowers.

Seven years ago, this meadow was a half-derelict piece of ground, part woodland, part ordinary lawn. It is now a perfect wildflower garden, a meadow thick with wild and antique plants from the first snowdrops and daffodils through scabious, knapweed, and meadowsweet. In the fall, there are silk masses of the grass-seed stems, the seed pods of the flowers, and the seasonal colors and fruit from the luxuriant surrounding hedges.

The site is ancient, and part of its boundary once formed the edge of a medieval forest. It is still marked with old hazels, once again being coppiced after centuries of neglect. In keeping with this romantic past, the owner wanted much of the garden devoted to wildflowers and wild animals, though with the addition of some old cultivated garden plants. It was also to be as "low maintenance" as possible. Now, in addition to its horticultural pleasures, the place is populated by glowworms and more than 22 species of butterfly, with newts, frogs, and several kinds of damselflies and dragonflies in the pool.

The meadow took shape quite slowly. The whole area was left alone for a season or two to see what was there, always the best policy with any new garden. Once the obvious omissions were clear, much of the area (heavy clay and mostly pretty wet) was double dug by hand. This is a perfectionist's job and took several more seasons; many gardeners starting a meadow simply kill off the existing vegetation with weedkiller, then sow their preferred seed mix the following spring.

The owner managed to harvest seed from other meadows nearby, collecting wildflowers like scabious, toadflax, vipers bugloss, and two knapweeds, though he also bought commercial seed of dropwort, salad burnet, and harebell. The seed was treated like that of any garden flower, being sown, pricked out into trays, then repotted into individual pots. The plants were semimature before they were planted out in the meadow itself.

The owner also wanted the meadow hedged and recreated an ancient mix of privet, dogwood, holly, bird cherry, hawthorn, wild roses, damsons, crab apples, euonymus, and hazel, all twined together with wild clematis and honeysuckle — exactly like the plantings used for kitchen-garden hedges in sixteenth-century Europe.

Main picture: Corn marigolds (*Chrysanthemum segetum*), wild cornflowers (*Centaurea cyanus*), and other meadow plants give a wonderful spread of color on either side of the mown path.

Above left and center: Mallow (*Malva moschata*) and thistle (*Cirsium vulgare*) make delightful and vigorous meadow plants.

Above right: Purple loose strife (*Lythrum salicaria*) gives tall spikes of brilliant color in damp parts of the meadow. Unfortunately this invasive perennial is outlawed in much of North America, where it displaces native wetland plants. If this is true of your area, try one of the many other species of *Lythrum* instead.

The whole area soon attracted some less garden-worthy wildlife, for a local deer herd found their way in and discovered that there was excellent grazing to be had among the hedges and the wild plants. They had, reluctantly, to be fenced out, and though the owner doesn't like the appearance of the deer fences, they do allow the meadow to grow. Given a few more seasons undisturbed by intruders, the hedge plants will screen the fence from view.

Now that the planting is complete, different areas have slightly different maintenance: one is mown fairly regularly (though buttercups and daisies flower regardless); another, the "spring meadow," is first cut in mid- to late July to let the cowslips and bluebells flower and seed. Even then, it is cut high enough to allow the cowslips to grow without losing their leaves. The area also has a scattering of wild daffodils (*Narcissus pseudonarcissus*), cultivated hybrids, and snow-

Above: Dandelion (*Taraxacum* spp.) and sorrel (*Rumex acetosa*) look as fine in fruit as in flower.

Right: A mown path makes it easy to view the flowers and the wildlife without having to wade through dense vegetation.

Right: Scabious (*Knautia arvensis*), bachelor's buttons (*Centaurea cyanus*), marjoram (*Origanum* spp.), and St.-John's-wort (*Hypericum* spp.) give rich hues and rich food for caterpillars.

drops (*Galanthus nivalis*, grown at least since medieval times), all of which thrive, though crocuses and tulips dwindle in the wet conditions.

The biggest area is devoted to the "hay meadows," where varying cutting regimes produce distinct flora. Two areas are cut in spring in alternate years, to allow butterfly larvae to survive and to encourage some plant species, like the southern marsh orchid (*Dactylorhiza praetermissa*), the handsome and poisonous belladonna (*Atropa belladonna*), wild marjoram, various catchflies and celandines, cuckoo flower, and the bronze-gold flowered *Pilosela aurantiaca,* introduced to Elizabethan gardens from southern Europe and now widely naturalized. Also grown here are wild forms of *Geranium pratense*, but many antique cultivated forms exist, too — look for the gorgeous seventeenth-century doubles.

At the edges of the meadowland, where it comes under the influence of the woodland boundary, the owners have planted foxgloves and the old red campion, often painted in medieval manuscripts and with a pretty double form in gardens since at least the 1600s — a good idea to copy. The meadow is close to the house, but paths are cut through the grass and mown frequently so that the flowers can be seen without tramping though long grass, new shoots, and seedlings. They make pleasant, unmuddy walks at any time of year.

All this comes in return for some extremely hard work at the outset, but thereafter it needs only one or two mowings a year and some occasional extra effort to make sure that stinging nettles and creeping thistle do not invade and crowd out the more valuable plants.

THE TOPIARY GARDEN

Topiary is an ancient garden art, whereby plants are made to form various green architectural shapes and their ability to flower and fruit is suppressed. It was immensely popular in the gardens of ancient Rome and has continued so almost ever since. Pliny the Younger (62–110 A.D.) described his Tuscan garden, parts of it filled with topiary figures made of clipped boxwood and one even imitating a Roman arena, with box seating and box inhabitants. Indeed, from Pliny to the Arts and Crafts movement almost 2,000 years later, there have been only slight changes in style, and the shape of much modern topiary would have been entirely familiar to Roman gardeners.

Topiary called for large numbers of gardeners to keep it controlled and shaped, and after the collapse of Roman wealth, there is little information about it until 1300, when Petrus Crescentius' *In commodorum ruralium,* the first modern treatise containing information on topiary, appeared in Europe. At first circulated only in manuscript, it was published in 1490, soon after the invention of book printing.

Opposite: Grand topiary, here in two variants of common yew at the famous garden of Levens Hall in Cumbria, in north-western England, gives powerful structure to the masses of bedding flowers (in the foreground, deep purple heliotrope). Above: Boxwood clipped into a ball is a style that would have been familiar to Roman gardeners and has hardly been out of favor since.

Above: The gardens of Barncluith, Strathclyde, in Scotland, built in the seventeenth century and currently being restored, are shown here in a nineteenth-century painting; the topiaries depicted may have been planted in the early nineteenth century.

Opposite: Neatly trimmed yew topiary in a Versailles tub, reminiscent of the style popular throughout Europe in the seventeenth century.

The Renaissance was founded on renewed interest in how ancient Greeks and Romans lived and thought, and on the desire to emulate their grandeur. Some great Renaissance families made determined efforts to outdo the achievements of the past, and their sumptuous villas had huge topiary gardens; the Rucellai Garden at Quaracchi in Florence had topiary spheres, porticos, temples, vases, urns, apes, donkeys, oxen, a bear, giant men and women, warriors, a harpy, philosophers, popes, and cardinals. The Medici Gardens rivaled them with elephants, a ship with sails, a hare with its ears up, an antlered deer, and more. The rich all over Europe copied the Italian grandees; Henry VIII had plenty of topiaries at Hampton Court, and Elizabeth I continued the show, with centaurs, servants with baskets, and so on.

Topiary remained vastly popular in the seventeenth century but reached its apogee in Holland, where there were boxwood and yew obelisks, pyramids, balls, triumphal arches, even whole theaters with stages and topiary scenery. Indeed, topiary is still popular in Dutch gardens, and the nurseries of Holland are the biggest suppliers of commercial topiary: along with globes and obelisks, it is possible to buy spirals and tiered "cheeses," though no antlered deer or centaurs.

The only serious dip in the popularity of topiary came at the beginning of the English landscape movement in the 1700s, when writers like Alexander Pope poked fun at the very idea of it. In North America, topiary seems to have been used, if modestly, in all grander seventeenth- and eighteenth-century gardens, and recreations of how it looked can be seen at Carter's Grove, Williamsburg, or Gunston Hall, Virginia, and at the Mayflower Society House, Plymouth, Massachusetts.

In the next century, fashionable gardeners changed tack once more, and by the 1870s and 1880s every new garden throughout Europe and America had at least one piece of topiary, and sometimes hundreds. By far the largest percentage of great topiary gardens dates from these decades and from the early part of the twentieth century: the great obelisks at Athelhampton in Dorset, or Waddesdon Manor, Buckinghamshire, which could hardly contain more topiary; the immense clipped hedges and domes at Old Westbury, Long Island, or at Oatlands in Virginia; and the vast scheme at Villandry on the banks of the Loire, as well as endless peacocks and cheese-stacks in village gardens throughout Europe.

Though demanding of time and patience, topiary is the most fantasy-filled part of gardening and has been of absorbing interest to all ranks of gardener. Certainly, for most antique gardens, or ones using antique plants, some form of topiary makes an exciting contrast to the foliage and color of the rest of the planting. Not only can it give a magical emphasis to the garden scenes of summer, it also provides something architectural to look at in winter, when most garden foliage and color have faded.

Topiary can make a glorious garden in its own right, whether you plan four topiary balls around a fountain in a tiny paved area or something grander, with obelisks or chessmen, or the more abstract geometric shapes sometimes found in ancient Italian gardens. Although many classical topiaries confine themselves to dark green boxwood or yew, try using different colors of whatever plant you use. This can alter the color balance of the whole garden and can sometimes be very dramatic. Victorian and modern gardeners have even used different species or varieties of a plant for different parts of the same topiary, perhaps having a golden yew peacock standing above a green yew base.

Topiary usually looks best in enclosed spaces, whether the enclosure is made of more hedging, stone, or brick, and even if the space is quite large. Enclosed, it does not have to compete with the outside landscape (or townscape), so its impact is much heightened, and the fantasy has no means of escape. Though you can make interesting topiary gardens in ten seasons or so, give some thought to longer-term plans. A walled garden in the Loire region of France, behind a tiny manor house built in 1634, was replanted in the early decades of the twentieth century: eight square beds edged with boxwood, yew pyramids at the corners, boxwood drums around the sundial, yew chessmen by the gates. It was planned as a recreation of what once might have been there, and now, sixty or so years later, nothing is more magical than wandering through the silent topiary garden on a misty moonlit fall evening, when the great shapes seem to have a mysterious life of their own.

Plants for Topiaries and Hedges

Classic positions for a single plant or a few topiary specimens are the junctions of paths, beside gates and doorways, by the sides of steps, or on the lawn. A row of them along a pathway can look splendid, and alternating them with unpruned trees, especially small varieties of fruit trees, can be immensely pleasing.

Let the plants grow as freely as they like and feed them well as you gradually cut them into shape. Alternatively, plan the shape in advance, make a framework outline of the final shape with bamboo stakes or sturdy wire and start clipping only once shoots get outside the notional frontier. Keep feeding young plants as well as possible, but stop once the topiary is fully formed.

How long does it take? Well, I've seen a handsome little peacock made from golden honeysuckle, about four seasons old. A decent boxwood ball 2 feet across might take four or five years, and a full-scale yew peacock ten to fifteen — though you could halve the time with something duller like privet. Whether it's a lot of work depends on how elaborate a shape you want. In general, to keep yew topiary in focus, clip it two or three times a season. Boxwood needs to be cut twice.

Below: Perfectly clipped balls of boxwood make a good counterpoint to a pretty statue draped with a strand or two of ivy, all backed by a handsome yew hedge.

Buxus BOXWOOD

Boxwood has been in gardens since gardening began, so it is not surprising that there are many variants. Common boxwood (*Buxus sempervirens*) itself has around thirty. For medium-sized hedges and topiaries, use either the basic species or the easily found 'Gold Tip,' whose young leaves have a yellow margin that fades as the leaves mature. Stems can grow 8in (20cm) a year. Boxwood can be used to make tiny hedges, about 6in (15cm) high, though better, if you want something so minute, is *B. sempervirens* 'Suffruticosa,' which was available in the seventeenth century.

Many nineteenth-century gardens have overgrown bushes of the various gold- and silver-variegated forms like *Buxus sempervirens* 'Aureovariegata' and *B. sempervirens* 'Argentea,' extremely attractive plants when left unpruned, with dense and glossy foliage. Those splashed and edged with silver seem less tolerant of neglect and drought, so are less suitable for potted topiaries. In seventeenth- and eighteenth-century New England, boxwood was the only topiary plant, which suggests that none of the specimens was large; boxwood can make fine globes up to a yard or more in diameter, cubes of the same scale, and obelisks up to about twice that.

If you want to grow several types in a hedge, try plain boxwood for the main hedge and a different variety where one path joins another.

Above: English holly (*Ilex aquifolium*) treated in a classical eighteenth-century manner as part of a garden enclosure, giving an architectural flourish even in deepest winter.

Ilex HOLLY

Holly has been connected with magic all over Europe and sacred to several northern pre-Christian goddesses. Pliny the Younger grew it to keep sorcerers at bay, a function that it retained into the eighteenth century. Though there are many species to choose from, *Ilex aquifolium*, English holly, clips well, though the fallen leaves are a menace to collect, and takes whatever shape you desire. In the eighteenth century it was commonly planted every couple of yards along a yew or boxwood hedge; the hedge remained clipped below and the holly was pruned to make a stemmed globe above it.

Even by the eighteenth century, there were dozens of variants, with spineless or enormously spined leaves; with berries in red, pink, yellow, or white; leaves variously marked in silver or gold; and so on. Though variegated-leaf forms are often seen pruned into domes and obelisks in informal gardens (there are some in my own), the basic green sort (*Ilex aquifolium*) really looks best.

Prune holly with shears. Hedge trimmers cut the leaves, and the cut margins brown. Most varieties can grow about 9in (23cm) in a season. American gardeners should check their location; Thomas Jefferson wanted hollies at Monticello but found them difficult to grow.

Laurus LAUREL, SWEET BAY

Laurus nobilis is the laurel of the ancients, used to crown poets and heroes and sacred to Apollo and Aesculapius, god of medicine. Its wonderfully aromatic leaves, smelling of cloves, nutmeg, and citrus, have always been admired. It was planted extensively throughout the Roman Empire, for religious as well as culinary reasons, and must have been brought to northern Europe during Roman times. It presumably survived reasonably well, though tough northern winters can easily kill it to the roots.

In medieval times, gardens were frequently centered on a well or fountain, or on a central tree. The tree was quite commonly a bay laurel, a traditional symbol of constancy and a popular evergreen. The garden mentioned in Chaucer's poem "Complaint of Fair Anelida and False Arcite" has both — a well "that stood under a laurer always green." Bay laurel must have been taken to North America with the early shipments of bulbs in the late seventeenth century and it was certainly grown there in nineteenth-century solaria.

As with holly, clip bay using shears: it's slower than electric clippers, but there won't be too many damaged leaves. If you want something for the kitchen, check the plant you are buying. Individuals differ in the strength of their savor; several seedlings in my yard, from the same tree in Rome, smell with differing intensities.

There is a reasonably attractive gold-leafed bay, at its most handsome in the spring, dingy thereafter. It is slightly hardier than the normal form.

Taxus YEW

The yew for the antique garden is *Taxus baccata*, found wild from Northern Europe to the Himalaya, and in North Africa, too. It is tough and extremely long-lived — many European gardens boast of having yew trees a thousand years old, though some of them may be much older still.

The basic species is best for both hedge and topiary. The horizontal nature of the branchlets, giving a dense side growth to tall topiaries, allows a marvelous velvety quality if kept in trim. The antique Irish yew (*Taxus baccata* 'Fastigiata') — it was discovered at the end of the eighteenth century — has branchlets that will grow vertically if allowed to do so. The plant is often used to make quick obelisks, but the structure of branching and branchlet gives a much less attractive finish, and mature specimens fall apart under snow. "Golden" variants (more a sickly yellow) appeared in the mid-nineteenth century.

Taxus baccata makes topiary easily, though the results can be overpowering when they are very large, so they are best suited to large gardens. Young plants, if well fed, can grow about a foot in a single season.

Opposite above: Cones of clipped bay laurel make emphatic statements among softer plantings; as bay's growth is fairly open, large-scale designs are easier to accomplish.

Opposite below: The classic yew, with its dense and hardy growth, is the most flexible topiary plant, making everything from perfectly plain and architectural hedges to playful designs like this.

Left: The low boxwood hedges that form this parterre and edge the garden paths give shape and structure to the planting.

Plants for Knots and Parterres

The form of the "knot garden," which began to appear in the early sixteenth century, varied slightly over the two centuries or so of its popularity, but even at its simplest, it was a complicated and almost always symmetrical pattern of low, interwoven hedges. Some knots were so dense that the "infill" areas were left empty or were graveled. As the fashion progressed, knots became more open; the areas between the hedging were used to plant small flowers like the smaller bulbs, daisies, pinks, violets, and strawberries. Taller plants, big enough to interfere with the view of the overall design — which was seen best from the house, the walk along enclosing walls, or from a garden pavilion — were grown elsewhere.

The "parterre" garden had many forms. They began to appear in the seventeenth century, when the fascination with mere pattern-making had relaxed and the emphasis shifted strongly to the cultivation of all the new flowers arriving in western Europe, for which there was not enough room in the old "knots." The word *parterre* merely denotes patterns of plantable earth cut into grass or box- or lavender-edged beds with gravel paths between, the better to allow examination of the flowers. Parterres after 1820 were often adventurously shaped; contemporary garden magazines ran letters from gardeners horrified by monstrous lapses of taste, having seen a caterpillar parterre or one with butterfly- or tadpole-shaped beds.

Above: Masses of lavender, here left unclipped and allowed to flower, fill a topiary garden with one of the most evocative of all garden perfumes.

Fagus BEECH

As an alternative to the plants described on pages 80–82, beech can make a glorious hedge around a parterre. Sheared twice a year, the common beech (*Fagus sylvatica*) can reach almost any height — 30 ft (9m) was once common, and examples can still be found. Such large hedges were used to make grand allées, as at Versailles, and sometimes trained into elaborate arches, as at Beloeil in Belgium, where beech, hornbeam, and linden have been used to create theatrical views in quite small spaces. These astonishing gardens, laid out in the early nineteenth century, are filled with canals and ponds, and the screens of tall greenery and the clusters of gloriettes (several times taller than the ones created by apple trees on p.130) are reflected in the still water.

But clipped beeches can also be used in more domestic spaces, to enclose flower or kitchen gardens, or even a wilderness, as at Courances in France. In more modest gardens, they were often pruned into big drums, sometimes on each side of the main gate. In the eighteenth century, there were popular white and yellow variegated sorts, both of which can still be found — as *Fagus sylvatica* 'Albomarginata' and *F. sylvatica* 'Luteovariegata,' respectively. The purple *F. sylvatica* 'Purpurea' and its many forms seem to be nineteenth-century varieties and are commonly seen in Victorian woodlands. They can be used for hedges, though to my mind they are far too heavy a color and are certainly not to everyone's taste. Bear in mind that beeches are shallow rooted, and tall hedges can dry the soil nearby.

Lavandula LAVENDER

A genus with several species from around the Mediterranean, the hardiest being *Lavandula angustifolia*, used by the Romans and brought by them to northern Europe. It was in every garden from at least the sixteenth century, yielding essential oils for perfumery and medicine, as well as pure garden pleasures. By late medieval times, it was used for low hedges in knot gardens and later in parterres. Though all these need clipping to keep them neat, the plants must have been allowed to flower (the young flowerheads are the "crop" for distillation), for a number of flower variants were grown by the late sixteenth century. Gerard grew the handsome white-flowered *Lavandula angustifolia* 'Alba.'

The usual blue lavender had crossed the Atlantic by the seventeenth century, though it needed protection from cold North American winters. A lovely pink-flowered variant called 'Rosea' or 'Hidcote Pink' is probably recent.

Rosmarinus ROSEMARY

Rosmarinus officinalis was used by the ancient Romans, who associated it with fidelity in love and thought and made garlands and coronets from it. They may have introduced it into Britain, where it is mentioned in eleventh-century manuscripts. Sir Thomas More wrote, in the early sixteenth century, "As for Rosemary, I lette it run all over my garden walls, not onlie because my bees love it, but because it is the herbe sacred to remembrance and therefore to friendship..." In the next century, the great English botanist John Parkinson grew three kinds, including the "gilded rosemary" (still around but much less good-looking than it sounds). Rosemary had reached North America by the seventeenth century; white- and pink-flowered sorts (*R. officinalis* 'Albus' and *R. officinalis* 'Majorca Pink,' respectively) were available in the nineteenth. In northern states, rosemary was treated as a greenhouse or conservatory plant.

All can be pruned to make excellent low hedges, say 2–3ft (60–90cm) high and deep. They are especially good as hedges in the rose garden; brushing against them as you sniff the roses adds a good bass note to the "bouquet" of the grander flowers, and also gives you something to smell when the roses are finished.

Santolina LAVENDER COTTON

Lavender cotton (*Santolina chamaecyparissus*) has been gardened all over Europe since at least 1548 and probably long before. Of Mediterranean origin, it is hardy in much of northern Europe but, even so, may have been rare in Britain; Gerard wrote that it was a smart plant, used "to border knots with." By the eighteenth century, it was used to edge walks.

Above: Rosemary, particularly the white-flowered form, clips well into hedges or low obelisks.

Below: Beech holds its warm, rusty brown foliage well into the gales of late winter, giving rich color to the topiary garden.

Left: Silvery santolina foliage, kept well trimmed, contrasts brilliantly with box hedging; the lower bed is filled with purple-leafed sage (*Salvia officinalis*).

A Topiary Garden

This garden, enclosed in a rectangle formed by barns and an old farmhouse, was created from what was almost a wilderness about 20 years ago. It has reached something approaching its present form only within the last few years and is slowly being developed and extended. Much of it is formal in conception, and there are topiary elements throughout; the small section shown here is made up almost entirely of topiary and heavily pruned apple trees.

Left: Boxwood is one of the most disciplined topiary plants, holding its clipped shape well. Trimmed and untrimmed plants side by side are neat without being regimental.

Below: Topiary doesn't have to be flamboyant. These carefully trimmed tables make their own subtle statement.

Owned by a famous interior designer, and with early planning and planting input from a well-known garden designer, the garden was intended to include many visual references to the owners' personal life, using Mediterranean and Middle and Far Eastern colors and forms (and occasional artifacts); ancient plants such as yew, myrtle, and boxwood; intense variations of light and shade; and ideas taken from gardens of seventeenth- and eighteenth-century Italy and France.

While the whole garden is an essay in the various uses of simple topiary forms, it is not in the least in the cottage-garden style. The garden uses topiary in ways that are both stylish and innovative. It is also steeped in visual imagination, from the contemplative "rill garden" — surrounded by dark yew hedges but with its central pebbled area enclosed by the artificial stream, the stones moistened just enough to ensure that they develop their most sumptuous colors — to marvelously planted pots and tubs.

The most immediate visual impact in the enclosed, almost cloisterish garden is given by the large rectangles of topiary, which were all, in the original plans, flat. The subtle low obelisks add a satisfying new twist to the look of the beds, powerfully framed as several of them are with double outer hedges.

Side sections of clipped boxwood form a grid or very simplified parterre, enclosing silvery clipped tufts of *Santolina chamaecyparissus* (*see p.85*), a fine contrast to the boxwood. The santolinas get several clippings through the season to discourage the garish yellow flowers.

Above: Topiary looks best in an enclosed space, whether the enclosing is done by yew hedging or the walls of the house. The area need not be small, but the impact of the planting is spoiled if the garden is allowed to blend into the outside world.

Left: Seen from the house, the enclosed space, filled with hedges of clipped boxwood, radiates order and serenity.

Right: As a counterpoint to the trim topiary, the farther part of the garden has apple trees pruned into dramatic shapes and irregular flagstones set into the grass beneath them.

Below: Flanking the fine bench are pots of *Acanthus spinosus* and clipped boxwood, symmetrically placed and drawing all the elements of the garden into a cohesive whole.

The boxwood itself is pruned two or three times during the year, using a double-sided hedge-trimmer. The small hedges that form the parterre are easy enough to maintain, but the flat tables of boxwood are more difficult to trim. The gardener simply wades in between the plants (which were originally arranged at 1ft/30cm intervals) and prunes each table with the help of a very accurate eye. Once the clippings have been roughly brushed off, the box topiary is easily combed back into its usual evenly close-knit surface.

This part of the garden has such a dramatic impact that it takes a moment for the visitor to see the farther garden, subtly closed off at the sides by yew buttresses and containing a theater troupe of apple trees, ten in all, arms raised, in double rows on each side of a flagstone path set flush with the grass. The neat cushion of greenery at the base of each is *Hebe rakaiensis*, lightly trimmed, and replaced every few seasons as it gets ragged enough to spoil the effect.

The apples are pruned as the beginnings of arches, while two groups of four at the front link up to make a pair of tiny gloriettes. The pruning means work both during the winter and also just past midsummer when the fruit has set, when excess foliage needs to be cleared away to let light to the fruit. The owner would like a much denser growth of branches, to provide heavier shade and stronger contrasts with the adjacent areas of the garden. If this is to happen, both grass and hebes will suffer and that section of the garden will need to be fundamentally rethought.

There are also some splendid contrasts between yew hedges (often 10–16ft/3–5m high), trimmed into architectural shapes, and grand foliage shrubs, left unpruned placed in front to give luxuriant effects. Viburnums, with magnificently pleated and veined leaves (*Viburnum rhytidophyllum* would substitute for the unobtainable one grown here), look particularly good, while the pea- and jade-green *Euphorbia characias* ssp. *wulfenii*, and less spectacular species, such as the variegated spiraeas and privets, are shown to great advantage. Elsewhere, domes of clipped boxwood sit at the front of beds, like theater footlights among borders billowing with antique roses and ivies.

Although they are not immediately apparent in the photographs, the garden contains some other interesting plantings. Along with the sumptuous pots of agapanthus, morning glories, and clipped myrtles, there are pots of *Acanthus spinosus* for summer foliage and flower spikes, but with the white-veined foliage of *Arum italicum* 'Pictum' for late winter and spring (the foliage dies away in summer). The combination is splendid.

Whatever the future developments of this garden, right now it has great presence. Bear in mind, though, that this sort of garden demands high maintenance. None of it is suitable for the time-pressed gardener.

Above: Domes of silvery *Santolina chamaecyparissus* push out from a grid of clipped box; these two plants have been used together since at least the sixteenth century.

THE ROSE GARDEN

The rose is among the oldest garden plants. There are references to rose oils, made from petals steeped in almond or olive oils, on clay tablets in the library at Nineveh, destroyed in 612 B.C. In Greece around 1100 B.C. rose oil was already used for perfume and to embalm the dead. The eastern Mediterranean region possesses many rose species, but *Rosa alba* and *R. gallica*, with their splendid scents, were probably the earliest to be cultivated.

Roses used for chaplets in Egyptian tombs dating from around 100 A.D. seem to be of a lovely hybrid that can still be grown: *Rosa sancta*, now called *R.* x *richardii*. In ancient Rome, roses were worn as wreaths for the head, to cool the brow and keep smell and disease at bay and were used as flavoring for food, in ointments and perfumes, and to bury the dead. The rich slept on pillows of rose petals, and the Emperor Heliogabalus is supposed to have suffocated some unfortunate dinner guests in vast mounds of petals dropped from the ceiling of a banqueting pavilion.

Opposite: The antique rose 'Blairii No. 2,' here combined with the more recent 'Fleurette' and a white clematis, makes a gorgeous perfumed cascade. Above: Shrub roses are tough and long-lived. This one, painted by Redouté for the Empress Josephine, seems to have vanished but may, like many others, still survive in old gardens.

Opposite: Redouté's painting of 'Rosa indica,' probably the variety now called 'Rosier des Indes'; it is still grown at Malmaison in France.

Below: An Edwardian rose garden in fall, with hybrid tea roses in flower.

Naturally, such demand for roses necessitated a huge cultivation of them, and the island of Rhodes was famous for its rose production. It is not clear how many roses the Romans grew, but they probably had the lovely *Rosa* 'Alba Maxima,' a double form of *R. gallica*, and *R. damascena*. Each of these went on to give rise to whole groups of varieties, which have survived through the centuries and still have many representatives today. However, as many nineteenth-century roses are complicated crosses between all these groups, they are very much obscured by the hundreds of intermediate types and in, many cases, it is impossible to ascribe a rose variety to any of them. This is no disadvantage, for even if it were possible, the classification tells the gardener very little about the qualities of any particular variety.

New roses gradually developed in the East and reached the West along trade routes or were brought back by Crusaders. In early medieval Persia, rose gardens feature in much poetry, especially that of Sa'di (1194–1296 A.D.), for whom a *Gulistan* is both a rose garden and a collection of poems. In Persia the

double yellow form of *Rosa hemisphaerica* (syn. *R. lutea*), which arrived in Britain in 1583, was often worn behind the ear, given as a present, or used for rose water. In the East, too, garden plants were collected, especially by Al Mu'tasim at Samarra in modern Iraq, where there was an extensive rose collection. Not surprisingly, roses figure strongly in garden descriptions in *The Arabian Nights*, whose stories were probably first collected in the twelfth century.

Roses grown in princely Europe, native species as well as ones treasured in ancient Rome, can be seen on the page borders of medieval psalters and prayer books. The species depicted is often the wild dog rose, but there are also red, pale pink, and white "cabbage" types. It was common in Italian and French gardens of the thirteenth and fourteenth centuries to find vines, roses, and water, reflecting the mathematics, geometry, symbolism, and philosophical ideas from the Near East. Boccaccio's *Decameron* describes the Villa Palmieri, where ladies sheltered from city plague. There, "with several walks and alleys long and spacious..., covered with spreading vines, whereon the grapes hung in copious clusters, and surrounded by white and red roses and jasmines...," the perfumes protected them from infection.

In Islamic Spain of the same period, roses were grown more imaginatively: at the Alcazar, there are still fragments of gloriettas left: arbors of cypress, twined with roses and jasmine, and used for dining. In northern India, Shah Jahan built the island garden of the Four Plane Trees, square, of half an acre and planted entirely with roses, stocks, marigolds, and vines — a planting easily copied today. It is an interesting color scheme, too, for he had something new: a rose called 'Dou Roux' with petals red on one side, copper yellow on the other, and so rare that feasts were planned for its blooming. It may have been *Rosa foetida* 'Bicolor,' still found and looking astonishing among modern marigolds.

A few medieval roses exist: the white *Rosa alba* in single and double forms, the red *R. gallica officinalis*, the dog rose (*R. canina*), the damask rose (*R. damascena*), and wild varieties such as the eglantine and *R. arvensis* (the only one that can be used as a climber). There were scarcely more Renaissance roses, and only nine garden sorts known in Holland or Britain by 1684. Widely admired were the cabbage roses belonging to the species *R.* x *centifolia*, which was soon to appear in many Dutch flower paintings and had probably been recently introduced from Constantinople, despite the fact that it was called the Provence rose. Genetic analysis suggests that it had a complex parentage, being descended from gallica, damask, dog, musk, and Phoenician roses. It, too, went on swiftly to establish a large number of subsidiary varieties.

Renaissance and seventeenth-century roses still in gardens include the charming *Rosa majalis*, commonly sold under its former name of *R. cinnamomea*; *R.* x *francofurtana*, now often found under the more recent name 'Empress Josephine'; the extraordinary *R. gallica* 'Versicolor,' often called *R. mundi*, which is striped and splashed in two shades of pink; the yellow *R. hemisphaerica*; the double-flowered *R. moschata*; and the very first American rose, *R. virginiana*.

Rose breeding took off in the 1780s, and every season saw new varieties. The introduction, in 1786, of the four Chinese floribunda-type roses, with their double flowering season, gave a great impetus to the rose industry. The crosses with European types gave rise to the hardy perpetual roses that were in flower from early summer to the first frosts. However, new sorts of roses continued to pour into Europe from America, India, and China, all used for breeding at once.

But the reverse happened too, to make the evolutionary story more complicated, and the new European varieties were being shipped, potted, to Carolina by 1732 and to China soon after. In their new homes, the roses were crossed with native species or into varieties that had arrived earlier.

By 1831 the colossal *General System of Gardening and Botany* by David Don, intended to list every plant grown in European gardens, named, but didn't describe, 211 Scotch roses, 54 damasks, 68 centifolias and 706 other roses probably belonging to the centifolia group — almost all with French names. Among these were the ones painted by Redouté, to capture the new glories for one of the most illustrious rose collectors, the Empress Josephine. Once she was married to Napoleon, and he was emperor, Josephine made their modest country retreat at Malmaison into a more fitting house. She seems to have been a genuinely keen and interested gardener and, falling in love with the new roses of the early nineteenth century, had assembled a world-famous collection by the time of her death in 1814.

Many Victorian roses survive: a number of the moss roses, the wonderful *Rosa* 'Souvenir de la Malmaison' (widely grown as a potted rose and still excellent though, alas, bred twenty years after Josephine's death), *R.* 'Gloire de Dijon,' *R.* 'Louise Odier,' and many others. Indeed, most of what we now call the "old" roses date from around 1850 or so and are of French provenance.

So exciting was the huge wave of introductions of new species and the breeding of new varieties that by 1800 many gardeners found that they had to have an area in their gardens entirely devoted to roses, in order to grow even a fraction of the marvelous innovations. By the middle of the century, the rose was probably the most fashionable flower of all.

The Victorian "rosarium" was the subject of debate: there were innumerable popular layouts, but one of the most common was to have a large circular bed divided into segments, one for each variety or group. Rose beds were often edged with half-hardy bedding plants, wallflowers, or tiny annuals, so that there was still something to look at once the main rose-flowering season was over. Vertical emphasis, where needed, could be given by trelliswork obelisks or arches, supporting new climbing varieties.

Some Victorian gardeners tried more daring ways of growing some of the weaker-stemmed bush roses, particularly the "basket of roses": circular or rectangular beds, often almost 6ft (1.8m) across, edged with wire basketwork 18in (45cm) tall. The earth inside was built up into a mound 2ft (60cm) high; a number of plants of one variety were put in, and as they grew, their stems were arched down and pinned to the earth. Eventually the desired effect — a tight and floriferous cone of roses — was formed.

Left: *Rosa virginiana*, the first of the American species to reach European gardens, arrived in the seventeenth century.

Shrub Roses

Most shrub roses have stems that are rarely more than 4½ft (1.5m) long, though *Rosa* 'Alba Maxima' can reach 6½ ft (2m) or more. Stems can be thin and floppy, making a spreading bush (as in *R.* 'Fantin Latour'), or thicker and upright, making the bush a good deal neater. Though the individual flowers of old shrub roses look perfect, some varieties can flop badly when the weight of the flowers is heaviest. It is worth placing low frames or rustic poles around each bush, with crossbars about 3ft (90cm) above ground level. This will keep the flowers well above other plants and let you see and smell them without too much effort.

The growth habit of many old roses means that it is easy to fill the ground beneath them with plants that appreciate the shade. Forms of the lesser periwinkle (*Vinca minor*), with their dark green, glossy leaves, make a good contrast, but if you want more diversity, include ferns, violas — especially the variants of *Viola cornuta*, which look wonderful with the sharp pinks of many old roses — ivies, some pale yellow foxgloves such as *Digitalis grandiflora* and *D. lutea*, and forms of *Campanula persicifolia*. That combination suits almost all old roses, even the darkest reds.

In very cold areas, not all roses are winter hardy, and some of the ones with Chinese and Indian species in their blood can be grown extremely well in the sunroom or greenhouse. Some of the Banksian roses are perfect for a wall in a cold climate, but try some of the shrub sorts in large pots, too. Many do well, producing much larger flowers under glass and flowering much earlier in the season. As many old roses have a short flowering season, growing them in pots means that they can also be placed out of the way, once flowering is finished.

Below: The soft pink flowers of *Rosa eglanteria* are followed by innumerable hips in a curious rusty red, but the real reason why the eglantine rose is essential in the garden lies in its leaves. Fresh ones have a powerful smell of apples cooking, especially strong just after rain, and that delicious perfume wafts around the yard.

Rosa 'Alba Maxima'

Sometimes called the Jacobite rose, this variety is certainly older than that cause, dating from at least the early seventeenth century. The Romans seem to have had a double *Rosa alba*, though it may have been the one called *R.* 'Semiplena.' All are disease free, tough, free flowering, and quite large. Plant them at the back of your border or on rustic tripods as the centerpieces to rosebeds.

Rosa eglanteria SWEET BRIER

This is the eglantine of Shakespeare and was probably grown in the ancient world. It was also one of the first European species to reach North America, when Dutch settlers planted seed in the yards of New Holland. They can hardly have wanted it for its flowers. However, it makes a marvelous hedge, either left to do as it will, in which case it needs 7–8ft (2.5–2.75m) of space, or roughly trimmed once or twice a season, when it rapidly makes an impenetrable barrier to all but the smallest creatures. In the eighteenth century, branches were commonly cut to put among other flowers in a bowl in the hall or parlor. If you need a rose hedge with more flowers, add some of the Scots briers, which are equally tough but have enchanting blooms, often exquisitely scented.

Rosa 'Empress Josephine'

Illustrated in 1820, this wonderful rose flowers in quantity, on a pretty, almost thornless bush. It is in fact older than the early nineteenth century: under its earlier name of *Rosa francofurtana*, it dates back at least to the 1580s and may be older even than that. Certainly the first one in British gardens was imported by John Tradescant the Elder, who collected it in Frankfurt in 1618, while accompanying an English ambassador journeying to Russia. The smell is good, too.

Left: Despite the name, *Rosa* 'Alba Maxima' is not quite white, for the young flowers have a pinkish glow, but it is perhaps one of the oldest doubles.

Above: *Rosa* 'Empress Josephine' is richly double and a marvelous sharp pink with deeper and stronger overtones.

Rosa foetida

Native to western Asia, this is the species that brought yellow — and susceptibility to black spot — into rose breeding. One form, *Rosa foetida* 'Bicolor,' has sumptuous single flowers, with petals that are colored copper-rose on the inside, mustardy yellow without. It has been known in the East since the twelfth century at least and in the West since the sixteenth. Persian writers called it the 'Dou Roux' rose, and feasts were held among its flowers. The form called 'Persiana' is all yellow and double; it reached Western Europe in 1837, to much acclaim.

Rosa × 'Harison's Yellow'

Though popularly called the Yellow Rose of Texas and sometimes given a species name of *Rosa* × *harisonii,* this is an 1830 cross between two Europeans — the ancient *R. foetida* and the little Scots rose *R. pimpinellifolia.* Making a bigger shrub than either, it seems to have been carried westward with the settlers' migration, so, unlike *R. pimpinellifolia,* it must "take" easily as cuttings. It is both handsome and hardy. Keep it away from pink old roses, planting it among irises, astrantias, the blue-green-flowered *Penstemon glabra* or *P.* 'Alice Hindley,' and *Alchemilla mollis.*

Rosa 'Ispahan'

Gorgeous, medium pink, packed with petals, marvelously perfumed, disease free, *Rosa* 'Ispahan' adds to its virtues by having a longer season than many other old roses. It has been known since 1832 and may have originated in the Persian city of its name. It looks good with *R.* 'Maiden's Blush' and *R.* 'Madame Hardy.' The growth is lax, and if unsupported, it makes a perfect rose to sprawl downwards over a low wall, beside a seating area, where you can easily sniff its delicious flowers. Underplant it with a mix of pale blue *Viola cornuta* and periwinkles.

Rosa 'Madame Hardy'

Pure white flowers are cooled still further by a tiny central boss of green, showing only when the magnificent flower is fully open. The perfume is rich but similarly cool. The whole plant, first recorded is 1832, exudes style, is a strong grower, and soon makes a fine large bush. Along with 'Madame Legras de St. Germain,' it is one of the most beautiful of the white roses. Grow it among pink lavenders, clumps of lungworts in its shade, and artemisias.

Rosa 'Madame Isaac Pereire'

Dating from 1881, this rose provides one of the great perfumes of the garden. Powerful stems and foliage make it easy to care for, and it has a scattering of blooms into fall. I grow it along a rustic fence by my kitchen garden, where valerians, white-fruited strawberries, and the black-flowered *Iris chrysographes* 'Black Knight' are also planted. However, it is powerful enough to be allowed to clamber up a wall or into an apple tree, where it can look beautiful. The banker to whom the immortal Madame Pereire was married eventually bankrupted himself.

Opposite above: *Rosa foetida* 'Persiana' — in the West since 1837 — is all yellow and delightfully double.

Opposite below: *Rosa* x 'Harison's Yellow' — the Yellow Rose of Texas — is in fact of nineteenth-century European origin.

Center: The green "eye" at the center of *Rosa* 'Madame Hardy' makes the flower seem even whiter. The perfume is also delicious.

Above right: The neat habit of *Rosa* 'Ispahan' means that it looks perfect by the path or terrace, and it won't flop to impede your walk.

Below right: The scent of *Rosa* 'Madame Isaac Periere' pours from large, blowsy flowers, which start out a strong cerise pink and age rapidly to a grayer tone.

Above: *Rosa mundi*, with its striped petals and luscious smell, shoulc be in every collection of antique roses.

Rosa 'Maiden's Blush'

'Maiden's Blush' is of tough constitution, though not of buxom stature. A lovely cottage rose, popular all over Europe, it is certainly medieval, perhaps older still, and looks good against a green background, half-engulfed by *Iris pseudacorus*, hostas, and forms of *Geranium pratense*.

Rosa mundi

With its wonderful, lightly double flowers splashed and streaked in two shades of sugar pink, this may be the rose from "Fair Rosamund's" bower (Rosamund was the mistress of the English king Henry II, and her rose is therefore medieval). However, Sir Thomas Hanmer, writing in the early

seventeenth century, says that it was a fairly recent "sport" (a variant that arises from a single bud on a growing bush) from the double *Rosa gallica*, to which an occasional branch of *R. mundi* still returns. Whatever its origins, it is heavily perfumed and makes a splendid low hedge, though it is susceptible to mildew.

Rosa 'Quatre Saisons'

If you miss roses after midsummer, grow this ancient variety, probably the "twice-flowering" rose used in the rites of Aphrodite in Greece in 1000 B.C. and later of Venus in Rome. Prettily pink and very scented, it has been in gardens ever since and is used as a parent of other long-flowering groups. It gives a second wave of flowers, not as strong as the first, in late summer and early fall. Not big, it should be grown in front of other bush roses and among 'Old English' lavender, where it looks enchanting. Its blood runs strong in the lovely 'Rose de Rescht,' an old Persian variety.

Rosa 'Souvenir de la Malmaison'

Soft pink, with the petals arranged in clearly marked quarters, this rose thrives in a hot place in the yard and does even better in a large pot in the sunroom. Being a bourbon — and so has fall-flowering *Rosa chinensis* as well as summer-flowering "Damask" in its veins — it has a reasonably long flowering season. It dates from 1843, but the climbing form is modern.

Above left: The form of *Rosa* 'Maiden's Blush' usually sold is 'Great Maiden's Blush,' shown here. It is a lovely, refreshingly perfumed, soft pink rose with gracefully curled petals.

Above right: *Rosa* 'Souvenir de la Malmaison,' sensual and sumptuous, is one of the most magnificent nineteenth-century roses, and deserves grand treatment in the garden.

Below left: Probably one of the most ancient cultivated roses, this variety is now called *Rosa* 'Quatre Saisons,' though it has, in fact, only two waves of flowers, the second in late summer.

CLIMBERS

The ramblers, often hybrids of *Rosa wichuraiana*, mainly date from the early twentieth century and are not old enough to be included here. They commonly produce long canes and flower on second-year canes, which are cut out once flowering is finished. The proper climbers produce canes, too, but they keep growing and flowering vigorously. Many climbers are very vigorous (*R. filipes* 'Kiftsgate' can easily grow 15–25ft (4.5–7.5m) in a season); they need to be planted with some forethought.

Victorian gardeners with exotic inclinations made "gloriettes," based on Arabic examples; many chose *Rosa* 'Alba Maxima,' which can be treated as a climber. When northern gardeners needed more light than was appropriate in southern Spain — where these arbors were designed to provide much-needed shade — some used groups of metal rose arches and surrounded the structure with lavenders or myrtles to get the lovely resiny perfume on hot afternoons. If you want to do something similar, the double musk rose is still the best, and if your climate won't allow *Jasminum officinale*, use honeysuckles instead.

Good places for climbing roses to sprawl are along pergolas (where the large ones can look marvelous), up walls, over sheds, and up trees. Many climbing roses are, or are bred from, what are in effect lianas. They are content to clamber up trees of substantial size, and nothing can look lovelier than an old pear or apple tree swathed with the flowers of *Rosa* 'Paul's Himalayan Musk,' an Ayrshire such as *R.* 'Janet B. Wood,' or the magnificent *R.* 'The Garland.'

Likewise, a vigorous rose climbing up a wall needs sturdy wires or hooks to hold it in place and will need regular thinning if you don't want it to

tear loose in the wind or collapse under the weight of snow. Roses that have fallen free are often an impenetrable tangle, and should be cut right back to a couple of good stems; they will soon grow back. The risk is worthwhile, for almost nothing looks more romantic than a house, summer house, or shed half-enveloped in roses.

Pergolas (or arbors) need to be strong enough to support the weight of vegetation, particularly if you want to combine your roses with wisteria, clematis, honeysuckle, or jasmine. They also need to be high enough to allow sufficient head room, so that the plants can hang from the beams, but not get in your way as you pass beneath. Although only a rich mixture of plants will produce a truly romantic effect, the tangle will prevent your doing any pruning, so you will need to do more radical work every five or six seasons.

Rosa 'Blairii No. 2'

On a north-facing wall, and with care taken to keep down mildew, this is one of the loveliest of the pink climbers — with flowers of perfect form, petals staying a deeper pink near the center, and a perfume that gives many gardeners the shivers. It provides a few flowers, too, later in summer and will remind you all over again about how superior old roses are in form, color, and scent to almost any of the "moderns," even with their tea-rose blood. If you can bear to harvest it, the flowers of this variety cut well, and the slightly pendant habit of the flowers works well in a vase. It was first grown in 1845. Don't grow it twined with anything else; let as much air as possible through the foliage. In some seasons you will need to spray it with fungicide. When it is in flower all the work will seem worthwhile.

Opposite: *Rosa* 'Paul's Himalayan Musk' is one of the most vigorous climbers, delicately scented and flowering freely in late summer.

Above: *Rosa* 'Blairii No. 2,' with its all-pervading perfume, is an ideal rose to go over a small arbor, shading a garden bench.

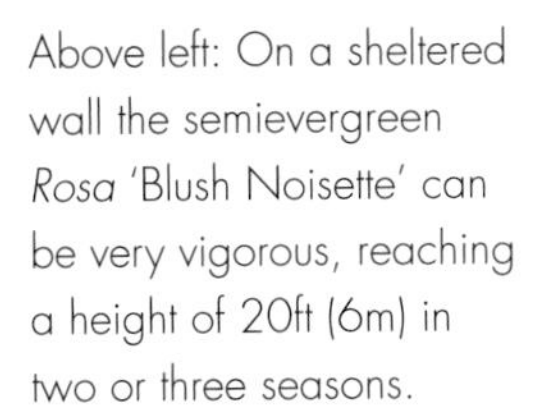

Above left: On a sheltered wall the semievergreen *Rosa* 'Blush Noisette' can be very vigorous, reaching a height of 20ft (6m) in two or three seasons.

Above right: The lightly scented *Rosa* 'Madame Alfred Carrière' continues flowering well into the fall. Fully hardy, it is a delight for northern gardens.

Rosa 'Blush Noisette'

Extremely vigorous in good soil and on a sunny wall, this rose also does well in less favorable circumstances and is disease resistant, too. From high summer it has big clusters of double, flattish, pale pink flowers that open from enchanting pinker buds and are perfumed with orange peel and spices. It can flower well into the fall and is easy to grow from cuttings. I grow it with white wisteria and the lovely old vine called 'Incana' or 'Dusty Miller.' The rose was first recorded in 1817.

Rosa 'Félicité et Perpétue'

Almost evergreen in mild gardens, this 1827 rose also flowers later than many climbers, for which it can be forgiven its lack of exciting perfume. Clusters of very double, very pretty, modestly pink flowers can cover the whole plant to dramatic effect. It is hardy, disease free, and good on a trellis or screen with grapevines (*Vitis vinifera* 'Purpurea' looks good) and the pale blue late-flowering *Clematis* × *jouiniana.*

Rosa 'Madame Alfred Carrière'

Famous on the cottage wall at Sissinghurst in Kent, this rose will grow in shade, too. Flowers are buff-white when young, true white later, rather tea-rose shaped, and with that delicious smell, too — a change from the headier perfumes of older roses. Only just antique, from 1879, it is a delight in the yard, with flowers into fall. I grow it as a backdrop to a planting of pink lavenders and the rose 'Empress Josephine' (*see p.97*).

Rosa 'The Garland'

Bred in 1835, this is a delightful and very vigorous rose. Red-purple stems and good foliage make it attractive even out of flower. Gertrude Jekyll liked it so much that she used it as swags on ropes strung along the veranda of her own house, Munstead Wood in Surrey. There can be no better recommendation. When I was once lucky enough to own an eighteeth-century summer house, I planted its walls with this rose, late Dutch honeysuckle, and *Jasminum stephanense.*

Above: *Rosa* 'The Garland' has airy clusters of small, amethyst-pink roses, fading as they age and slightly but spicily perfumed.

Left: *Rosa* 'Félicité et Perpétue' is another vigorous climber that copes well with the rigors of the fall. Many of what we now call "old" roses are of French provenance; this one is named after two otherwise forgotten sisters.

Right: Against a cottage wall, climbing roses make an attractive planting, with *Euphorbia characias* and sweet rocket (*Hesperis matronalis*) beneath.

Below: A single red form of *Rosa rugosa*, growing among cow parsley in a wild part of the garden.

A Garden for Old Roses

This rambling manor house must always have had roses of some sort planted around it, perhaps *Rosa gallica* 'Officinalis' and *R. mundi*, or later *R.* 'Celestial' in pale pink and the tiny deep pink *R.* 'De Meaux.' They would have been part of the general planting, but in the early 1900s the house was given a rose garden devoted to their culture alone. Set a short walk from the house, it is walled off from the rest of the garden, giving it a sense of retreat and privacy.

Left: The walk between rose arches is paved with brick laid to accentuate the length of the view.

Above: 'New Dawn' flowers late and does well in cold gardens.

Below: Not quite antique, 'Seagull' has a long heritage, and its beautiful fragrance earns it a place in any rose garden.

Perhaps the garden once had the wooden trellis-work arches and arbors that were fashionable in the seventeenth century, but they would have been smothered in musk roses, the uprights perhaps covered with *Rosa* 'Alba Maxima.' The present arbor is romantically swathed in dense growths of 'Seagull' and 'New Dawn.' Though 'New Dawn' dates only from the years in which this garden was built, it is a very attractive, easy, and constant-flowering derivative of 'Dr. W. Van Fleet,' an American hybrid raised in 1910.

Alternate arches are planted with the gorgeous and vigorous 'Seagull.' It dates from 1909, so is fast becoming an antique. It is an historic cross, though, one parent being the ancient Chinese species (and favored garden plant) *Rosa multiflora*. Introduced to Europe at the end of the eighteenth century, *R. multiflora* is a rose worth having in its own right, with huge clusters of fragrant white flowers. 'Seagull's' other parent was 'General Jacqueminot,' a gorgeous hybrid perpetual of 1852, which is still available.

'Seagull' is strongly perfumed, and though it blooms for a single season, producing only a few flowers later in the year, even then it will perfume the whole yard and beyond. If you want to plant your own rose arch, you should be able to achieve the same pitch of luxuriance in five or six good growing seasons.

Other roses in this part of the garden are grown with their stems twined around metal obelisks (though some blend even more delicately with structures built from rustic poles). The brightest is the flaming scarlet *Rosa* 'Excelsa,' from 1909. When in flower, sometimes twice a year, it possesses a lot of the characteristics of the Chinese *R. wichuraiana*, while the clustered flower buds come from *R. multiflora*. Although not quite an antique itself, 'Excelsa' therefore has links to two antique climbers. If you like the idea, try it with one of the antique climbers described on pp.102–105. Be warned, though, that only modern roses will give you quite this shade of red.

Elsewhere in the garden, shrub rose bushes have a topiary background, the dark green of the yew hedges making a splendid foil to the wild growth and the sumptuous colors of the roses. It is an effect well worth trying in a much smaller garden, either with topiary elements giving form, summer and winter, to the rose garden or with the structure being supplied by a backing fence or wall covered with the dark green foliage of ivy.

In yet another part of the garden, roses are grown in an entirely informal way, almost as if they were wild bushes, left to make their own way, either as specimens on a neatly mown lawn or in long grass. They can look equally romantic in either setting, so an informal rose garden could be bordered by — or even blend into — a meadow garden or kitchen garden.

Especially good for this treatment, and well represented in this garden, are some of the rugosa roses. These developed from the Japanese sand-dune species, which were first in Europe in the 1860s and used soon after for hybridization. 'Mrs. Anthony Waterer' is a good one dating from 1898, but in this garden the dark and sumptuous 'Roseraie de l'Haÿ' is from 1901, the paler 'Delicata' from 1898, and the handsome double white 'Blanche Double de Coubert' 1892. All have long flowering seasons, do not seem to be troubled by pests, and never suffer from black spot. They have strong perfumes, and as they also flower well in the shade of walls and buildings, they make excellent antiques for urban rose gardens.

However, no manor-house garden restricts all its roses to a special rose garden. Elsewhere in this garden, old roses are planted against the walls of

both the service cottages and the house itself. Entirely simple, some of the plantings by the windows are especially attractive. The walls hold 'Zéphirine Drouhin' and 'Madame Alfred Carrière,' both easy and beautiful, with a cottage-garden planting beneath of annual poppies, *Artemisia absinthium* 'Lambrook Silver,' the perfumed sweet rocket (*Hesperis matronalis*), and the long-lasting greenery of *Euphorbia characias* ssp. *wulfenii.*

Other parts of these beds contain eye-catching plants such as the silvery-leafed and sky-blue-flowered *Teucrium fruticans* and the gawky but pretty Victorian verbena, *Verbena patagonica* (syn. *V. bonariensis*). To complement the generous clumps of wooly-leafed *Stachys byzantina*, try adding an occasional bush of *Rosa* 'Empress Josephine' and a cluster of *R.* 'Ispahan' by the front or back door. On the wall, a less florid rose to try would be the delightful *R.* 'Blairii No. 2.'

Opposite: Wilder parts of the garden are fine places to grow species roses, like this yellow variant of the Carolina rose.

Left: Evening light makes the rose garden even more romantic.

Above: 'Ethel,' a popular, vigorous, and easy climber from the early 1900s.

Below: A fairly modern floribunda rose, of a shade of red not found in antique roses, but here bringing a welcome contrast to the whites and pale pinks.

The Kitchen Garden

This is the most ancient kind of garden, and many kitchen-garden plants appear in creation myths: palm, quince, grape vine, and apple, have all been seen as the cosmic Tree of Life (or Knowledge), growing atop a sacred mount, or in the Garden of Eden and are often associated in some way with snakes. Many, too, have been associated with magic and ritual into quite modern times. Kitchen gardens have also been places in which to proclaim status; in 1000 B.C., a Babylonian king wrote a catalog of the vegetables, fruit, and medicinal herbs in his gardens, and a few centuries later Sennacherib, King of Assyria, thought his so important that he built an aqueduct to carry water to its fruit trees.

Though royal gardens were on colossal scales, Homer describes a somewhat more domestic one: the gardens of Alcinous, an enclosed garden of about 4 acres, hedged, with fruit trees planted in rows (the first mention of a "quincunx," a grid planting that remained in use in the West well into the 1700s).

Opposite: Vegetables, herbs, and flowers such as marigolds and nasturtiums blend easily with the topiary boxwood and formal paths in this delightful potager. Above: A nineteenth-century turnip from the German Benary seed catalog. Use old varieties for the varying shades of their leaves. The young foliage and flower shoots are delicious.

In ancient Greece, too, the landscape seems to have been dotted with sacred gardens: an inscription at Thasos describes the lease of a garden devoted to Heracles outside the town, which was to provide figs, hazels, and myrtles and boasted porticos and banqueting rooms.

Gardens of the Roman Empire are better known. Recent excavations of Pompeian houses show that the courtyards were often filled with figs, cherries, apples, and pears, probably pruned high so flowers and vegetables could be grown beneath. Some trees had vines trained up the stems. Roman gardens were still places for the gods, not only those of the household, but the god of the garden, Priapus.

After the fall of Rome, the traditions of kitchen gardening continued in monasteries and princely gardens, and probably lower down the social scale. Many kept in at least distant touch with the developing traditions of the Orient, and soon influences from ever farther east were reaching western Europe; in 1404, the court of Tamburlaine, at Samarkand, was visited by the Castilian ambassador. Invited to various receptions in gardens, he was even lodged in one and described it in disbelieving terms: "We found it enclosed by a high wall... and within it is full of fruit trees of all kinds save only limes and citrons...." He eventually saw Tamburlaine, seated on a dais in front of an immense fountain, upon whose waters floated scarlet apples.

A century later, the Mughal Emperor Babur wrote his memoirs, including details of many of his most beloved gardens. These, illustrated with paintings commissioned by his son, show watercourses, vegetable beds, and bountiful fruit trees to provide shade for banquets.

As the Renaissance moved out from Italy a century or so later, a renewed interest in the diversity of plants, and most especially of fruit trees, spread. These were eagerly collected as badges of sophistication — intellectual and gastronomic. In England, Henry VIII and his gardeners are often thought of as having introduced all sorts of new crops to Britain, from Turk's head pumpkins (brought from Constantinople, though introduced there from central and southern North America), new kinds of greengage, spinach, the first apricot (at least since Roman times), apples, and others. In late sixteenth-century France, where the first parterres were beginning to evolve, collections of exotic fruit and vegetables had already been formed at the château of Blois. The Queen's garden had vegetables planted inside the new boxwood compartments of the great parterres. Farther north, Renaissance attitudes gradually enlivened the table; seventeenth-century Scottish and Scandinavian grandees had walls with trained fruit, sometimes terracing the steep site originally chosen for defensive reasons, sometimes using new walls enclosing the front courtyard, so that grand plants could be seen by all visitors. Among other exotica, they imported lemon trees from Italy and cauliflower seed from Crete.

Not only were Europe and the Near East being explored for new delicacies; the first American crops reached European shores in the late 1400s, but many more arrived in the succeeding century. These crops (the tomato, the potato, eggplant, peppers, maize, squash, and pumpkin) all made a colossal impact on European gardens and kitchens, as well as on the kitchens of the Near East and North Africa. Many of the new crops were so delicious, and so high yielding, that ancient native crops, especially among the root vegetables, began to drop out of use.

The show aspects of vegetable and fruit gardens began to wane just into the eighteenth century, when French influence was increasingly in the ascendant. In western Europe, the kitchen garden, though still an important part of any large garden, became a separate entity, often a walled enclosure at some distance from the main house. Many were extremely grand.

Extraordinary things were attempted in these gardens; the novelist Daniel Defoe found oranges being grown outdoors on an English estate, at least 500 miles (800km) north of where this is conventionally done: "...and [they] have moving houses [of glass and timber] to cover them in winter; they are loaded with fruit in the summer, and the gardeners told us, they have stood in the ground where they now grow above eighty years...."

Nineteenth-century ingenuity in the kitchen garden went on to produce endless new varieties of important crops, as the scientific aspects of hybridization developed. Ironically, the same century saw the growth of high-efficiency farming as well as a vast increase in the size of towns and cities. Both meant that most people could buy cheaply varieties bred for farm methods, rather than varieties bred for the more individual needs of gardening enthusiasts. Local varieties, some historic, some modern, became extinct in huge numbers.

Nevertheless, growing the surviving antiques of the kitchen garden can give the gardener, especially of fruit species, crops more delicious and more beautiful than anything that can be bought. Though a number of antique varieties can be found in commerce if you are prepared to do some hunting, many countries are now forming national collections, and it is often possible to arrange to have grafting material from trees that interest you. Examples are the Brogdale Trust, Faversham, England, with huge collections in all top fruit species, and one devoted to peaches in the Tarn region of France. All these collections run tasting, grafting, and other courses.

Opposite: This nineteenth-century painting from the Benary catalog, shows the mix of productive plants and ornamental flowers that characterized the traditional kitchen garden.

Below: Ornamental kitchen gardens always need vegetables with attractive foliage and stems. Ruby chard, shown here, is easy, but also look for exotic-leaved basils, sorrels, and brassicas.

FRUIT TREES AND ORCHARD GARDENS

If your property has room for a full-scale orchard, and you want to grow antique fruit trees, plant them as a quincunx, an arrangement of trees in use since ancient Greece and into modern times. The simplest layout had trees planted at the notional corners of a square, usually 15–25ft (4.5–7.5m) on the side, the grid generally running north/south and east/west, with the planting as large as the owner had room for. A more complex and rather more interesting one had a fifth tree in the center of each square.

If you can, give orchards some sort of enclosure. It was a moat and palisade in the grandest medieval examples, but a hedge of hawthorn or of mixed native trees and bushes, with simple hedge-plant fruits, may be more practical for most contemporary gardeners. Inner paths were often hedged with gooseberry or common barberry (*Berberis vulgaris* is the only one to use, though not in the U.S., where upright forms of blueberry were often substituted). The trees themselves had an underplanting of grass, scattered with sweet violets, wild strawberries, spring bulbs, and even an occasional rose bush (try *Rosa* 'Alba Maxima'). Seventeenth-century orchards often had their paths centered on a statue or an urn or even a small summerhouse — still a perfect centerpiece. If you prefer informality and want to plant the orchard as wilderness, keep the trees at least 15ft (4.5m) apart and mow circulation paths between them.

In smaller yards, just two or three trees will give great pleasure, but remember that, for the best crop, apples, pears, plums, and hazels all need a suitable "pollinator" combination of two different varieties. In tiny yards, try espalier-pruned trees against the house or a shed wall, or the single-stemmed cordons. Pruning does take time in winter and high summer, but the results look good and the crops can be heavy. If space is really limited, stick to apples and pears.

Malus APPLE

One of the most beautiful of all fruit trees, the apple probably originated in the Caucasus as a hybrid between a number of local species of *Malus*. Primitive sorts can be found all over the region between Turkey and India. The Romans had many types, and orchards were extremely profitable. In western Europe, apples gradually developed huge numbers of local varieties, many of which survived into the seventeenth century. Groups like the rennets, pippins, pearmains, and russets are all from this date, though breeding within them has continued into this century. Many were in America by the seventeenth century and have given rise to hundreds of American varieties.

Good older apples include, from before 1600, 'Nonpareil,' 'Golden Pippin,' 'Calville Blanc,' or 'Rouge d'Hiver'; 'Ribston Pippin' of about 1707; 'Ashmead's Kernel' from the late eighteenth century (a vigorous tree with aromatic fruit); and 'Cox's Orange Pippin' from around 1830.

Mespilus MEDLAR

A native of Europe, the medlar was apparently ignored by the Romans, but every Renaissance writer knew it. By then there were many recipes for its use in the kitchen. Fruits were harvested when hard if they were to be kept into midwinter or left on the tree until ripe (or "bletted"). The tree has large flowers in spring, a picturesque shape, and magnificent fall color. In the seventeenth century, an ornamental free-standing tree or two might have been planted in the kitchen garden.

Opposite: Apples, including 'Devonshire Quarrenden' (from before 1700), pruned as single-stem cordons and supported by metal arches.

Above: 'Court Plat Pendu' probably dates from well before the sixteenth century. The fruits, small and russet yellow, make delicious winter eating.

Below: *Mespilus* 'Nottingham,' with fruit like small, foreshortened pears in a fine amber brown. It has been known since the eighteenth century and is probably older.

Above left: *Vitis* 'Black Hamburgh' is good on walls or trained up the inside of a greenhouse. It fruits prolifically and is delicious.

Above center: The gooseberry 'Whinham's Industry' shows how dark ripe fruits of some varieties can become; others ripen almost black. All make delicious dessert fruit.

Above right: Morellos are the only cherries that take easily to pruning; they can be trained as large fans and will fruit on a shady wall.

Prunus CHERRY

Sweet cherries originated in Greece or Turkey. After Lucullus carried some back to Rome in triumph, they spread rapidly through the Empire: the 'Kentish Red' (a type of morello cherry, sharp flavored, but with a larger fruit) may date from Roman times. The variety called 'Noble,' still available, was imported to Britain from Belgium by Tradescant the Elder. The earliest to crop is the pale 'May Duke,' also seventeenth century. In eighteenth- and nineteenth-century orchards, trees were often underplanted with strawberries and currants; try that in yours. Edible cherries are as marvelous in flower as their ornamental relatives.

Prunus PLUM

Domesticated in Asia around 1000 B.C., long after the more refined damson, plums have been grown continuously since then. They reached North America in the 1500s. Each continent has its own distinct varieties, with Chinese ones especially distinctive. 'Coe's Golden Drop,' from the late 1700s, is a delicious British one to grow. The most successful of all must be 'Victoria,' now grown worldwide; it seems to have originated in 1840.

Prunus persica PEACH

Though called *Prunus persica*, the peach was first domesticated in China, where it was grown for its large pink flowers as well as the delicious fruit. It took many centuries to reach Europe; Pliny calls it "recently introduced," but soon many sorts were grown all over the Roman empire. There were red- and white-skinned peaches in Europe in the early 1500s, and the fruit reached America by 1565. There it became so common in some states that it was used to feed pigs or distilled to make brandy.

The varieties 'Rochester' and 'Peregrine' are antique and delicious. But look for the 'Pèche de vigne,' a variety that comes true from seed, velvety yellow and succulent, and was used in ancient and modern Italy as a prop for grapevines.

Pyrus PEAR

Originating in prehistoric western Asia, the pear was a "fruit of the gods" in Greece of 1000 B.C. The Romans had dozens of sorts, some for winter eating, and more than 1,000 sorts exist today. For antique gardens, look for the delicious 'Jargonelle,' 'Catillac,' and 'Bergamotte' of 1600. The fruit of all three is small, with the first making late summer eating, the others not softening until winter, so store them carefully. Trees produce abundant crops, so pickled pears and chutneys are other good uses. There are also large numbers of extant nineteenth-century varieties.

Ribes x *uva-crispa var. reclinatum* GOOSEBERRY

A European native, this was used by both the Greeks and Romans and was an ingredient of sauces in medieval Europe; new French sorts were being imported to Britain in the 1270s. Five hundred years later, there were so many enthusiastic growers that "gooseberry clubs" sprang up, and there were huge numbers of varieties. Sweet sorts remain most popular in Britain and North America, though tart ones are essential in France and Germany, in sauces for oily fish and meat.

Vitis GRAPEVINE

Native to the eastern Mediterranean, grapes were grown in the Middle East by 4000 B.C., and there are Egyptian manuals on viticulture from 1,000 years later. Early Romans treated wine warily, but late Romans had vineyards all over the Empire, and wine and raisins were important commodities (Greek wines were best, and Pliny drank some that was 200 years old). Norman Britain had many vineyards, as did great estates 700 years later. Antiques can be very old: 'Muscat Frontignac' flourished in ancient Athens; the easy-to-grow 'Black Hamburgh' is probably sixteenth century. American settlers used native species, but soon began to hybridize these with European cultivars.

Above left: Peaches are easily trained against walls for early crops. Even in cold gardens with late springs, they can fruit well if protected.

Above center: *Prunus* 'Coe's Drop': a delicious, juicy, eighteenth-century plum variety.

Above right: *Pyrus* 'William's Bon Chrétien' is a nineteenth-century representative of a group of autumn-ripening pears, the earliest of which were grown in France in the sixteenth century.

The Vegetable Garden

In the tiniest yard, where there is perhaps space for only a single fruit tree and a few rows of lettuces or vegetables, there is still room for something antique. For the tree, plant an old local variety of apple or peach, or a mulberry or medlar. Beneath it grow a row of skirret (*Sium sisarum*, with sweet parsnip-like roots) or of white Italian cucumbers and a pinch of antique radishes.

If you have more space, why not plant a potager? The combination of formal beds with vegetables and fruit can be quite as productive as the stolid rows of the conventional vegetable patch, and though proper crop rotation needs more planning than usual, a well-managed potager gives you somewhere pleasant to sit while you survey your growing crops. Providing there is room for an attractive formal arrangement of beds, there are endless visual possibilities with lettuces and vegetables. Try some of the red- or tan-leafed chicories or lettuces, the scarlet-veined ruby chard, beans with golden pods, or the brilliantly colored and edible flowers of antique varieties of borage, nasturtium, and marigold.

For the more permanent parts of the design, make gooseberries, currants, and all the tree fruits part of the decorative scheme, either by pruning them carefully into the desired shapes or by planting them in big pots or tubs.

Below: Onions, with their many variations in the shape and skin color of the bulbs, can make decorative as well as useful additions to the kitchen garden.

Allium cepa ONION

Probably grown in the earliest known gardens, in Mesopotamia, the onion was subsequently taken up by gardeners in ancient Egypt, Greece, and Rome and then given to the rest of the world. Oddly, Renaissance gardeners in Europe thought that it had all sorts of dangerous properties, and it wasn't even a cottage-garden plant until the nineteenth century. There are huge numbers of varieties (hence the expression "knowing your onions") — from the potato onions with their clusters of subterranean bulbs and the Egyptian onion with groups of bulblets instead of flowers, to scallions (once used as an aphrodisiac), spring onions, and many more. The mild-flavored and long-keeping 'Bedfordshire Champion' and silvery salad 'White Lisbon' are both early nineteenth century. The red-skinned 'Red Wethersfield' is an American variety of about the same date.

Welsh onions belong to a different species (*Allium fistulosum*), are ancient, and have been gardened in Europe since at least 1629.

Allium porrum LEEK

Though leeks have been the badge of Wales since the Dark Ages, the earliest record dates from Egypt in 3000 B.C.; even in first-century Rome, gourmets thought that the best ones came from there. The Emperor Nero imported them, thinking that they improved his voice. A much more varied crop then than now, forms were once used for their foliage, others for side bulbs and offsets. The difference between leek varieties is not large, so even modern ones won't look out of place in antique gardens. The heavy-cropping and very hardy 'Musselburgh' dates from around 1834 and was grown in North America soon after.

Amaranthus AMARANTHUS

This is a popular North American crop; many species give edible foliage and seeds. *Amaranthus hypochondriacus* can have remarkably ornamental foliage; the golden-leafed form is a handsome addition to the potager. But there are maroon, scarlet, and speckled variants, yielding brilliant dyes used by Native Americans; the leaves can also be steamed like spinach. Both this and *A. leucocarpus* are important grain crops.

Though amaranthuses reached Europe by the late sixteenth century, they have not really strayed much outside the flower garden, where forms like 'Prince's Feather' were much liked.

Above: *Amaranthus hypochondriacus* is a spectacular plant even in its basic form, but there are still more dramatic varieties with scarlet and golden foliage.

Left: Leeks, winter hardy and cropped throughout that season, give welcome structural emphasis to the kitchen garden when most of the ground is bare.

Above left: Fennel makes large and perennial plants that can be used to give scale to the kitchen plot, as well as flavor to salads and baked fish.

Below left: Artichokes, with their handsome foliage, are essential ingredients in the antique potager.

Below right: Savoy cabbages, hardy in all but the coldest gardens, are delicious and decorative.

Brassica CABBAGE

This is a huge group, probably domesticated in Asia Minor. The Greeks had a kale by 600 B.C., and the Romans grew a head-forming cabbage that they thought was bad for their eyesight. Red cabbages were grown in Germany by 1150, and three sorts of savoy were known by the sixteenth century. Many sorts have vanished, including the ancient "musk cabbage." One early variety, 'Couve Tronchuda,' with its open growth and fleshy pale midribs, is still used to make a special soup in Portugal. It tastes best after heavy frost. Other survivors include 'Early Jersey Wakefield' sorts of around 1880 (grown in American gardens by 1885). 'Dwarf Green Curled' savoy and 'Winningstadt,' a good variety for salads, both of 1859; and 'Wheelers Imperial,' with solid cones of gray-green foliage, from 1849. Perhaps the oldest is 'Premium Late Flat Dutch,' from before 1700.

Cucurbita PUMPKIN, SQUASH

This American crop covers several species of the genus. All were probably domesticated between 10,000 and 6000 B.C., originally for the seeds, roasted as they are today in the Middle East, rather than for the delicious flesh of the fruit itself. The marrows are most familiar in Europe, whether harvested young or left to mature. They crop in cooler climates than many pumpkins. Antiques include the probably ancient 'Vegetable Marrow' and 'Boston Marrow.' Pumpkins and squash are both better to eat and to look at.

As the marvelous fruits store well, they must have reached Europe very soon after Columbus landed and spread around the Mediterranean region soon after. They may have reached northern Europe from Turkey, hence the name 'Turk's Cap' ('Turk's Turban' in the U.S., where there seems to be no native name) sometime in the sixteenth century. Antique varieties such as 'Green Striped Cushaw' and 'Chestnut,' common in the U.S., are still unusual in Europe, though modern variants are becoming more familiar.

Cynara ARTICHOKE

The artichoke was admired by the ancient Greeks. In Rome, Pliny the Elder found that it was the most expensive vegetable in the markets, with crops imported from as far afield as North Africa. In medieval Europe, it was vastly popular and thought to be aphrodisiac. American gardens had artichokes by the sixteenth century. Plants grown from modern seed will cover the entire range of variation; alternatively, in both North America and

Europe, look for 'Green Globe' of 1828. Its big scales are not too heavily spined and have large quantities of flesh at their bases.

Foeniculum FENNEL

Fennel has been cultivated since earliest times, and Roman cooks had several forms from which to choose: "sweet" fennel in many variants (there were special sorts to provide roots, leaf, "bulbs," stems, and foliage), as well as "bitter" fennel (used principally to flavor alcohol).

Of the "sweet" fennels, some root varieties still available in Italy have roots tender all the way through like parsnips and are delicious. Particularly in the south, forms are also grown with juicy stems that do not become woody, which are sliced into salads. 'Zefa Fino,' grown in America since 1890, is an old 'Florence' fennel, a form in which the lowest parts of the leaf stalks are succulent and wrap around one another to make a sort of bulb. Of the herb (bitter) fennels, try the fine one with purple-bronze foliage and stems.

Above: 'Turk's Turban' pumpkins have very decorative fruits that last well when harvested. The deep orange flesh is richly flavored and delicious when baked.

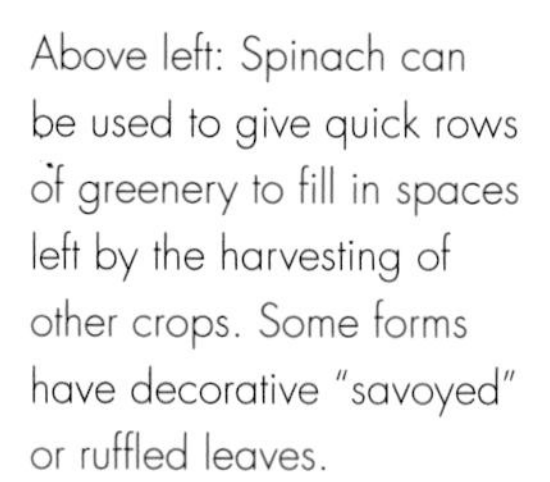

Above left: Spinach can be used to give quick rows of greenery to fill in spaces left by the harvesting of other crops. Some forms have decorative "savoyed" or ruffled leaves.

Above right: The runner bean 'Painted Lady' (of the early seventeenth century) makes a handsome climber and crops heavily, too.

Right: The wild tomato is not too distant from the 'Gardener's Delight' shown here, which looks delightful scrambling over an arbor or up a tepee of poles.

Opposite: The pea 'Kelvedon Wonder,' a form of one of the most ancient garden crops.

Lycopersicon TOMATO

Strictly a fruit, tomatoes are included here because of their widespread use in salads and main dishes. Probably first seen by Europeans in Mexico in the sixteenth century, they were thought to be poisonous. Why what was so obviously a crop plant for the Native Americans should be regarded as poisonous in Europe is not clear, though the flower and fruit do resemble a number of very poisonous European members of the family.

In Europe, tomatoes first entered the kitchen in Italy, and the habit soon spread around the Mediterranean. They were rarely eaten in cautious northern Europe until the nineteenth and even early twentieth century.

Early introductions included red, yellow, and white forms, with the red in two varieties: small and round or large, flat, and lobed. This last sort was thought to have the best flavor (it still has) and is now referred to as part of the 'Marmande' group — the one seen most commonly in French country markets. Many sorts of cherry tomato were in widespread cultivation in Central America by the time of the Spanish invasion.

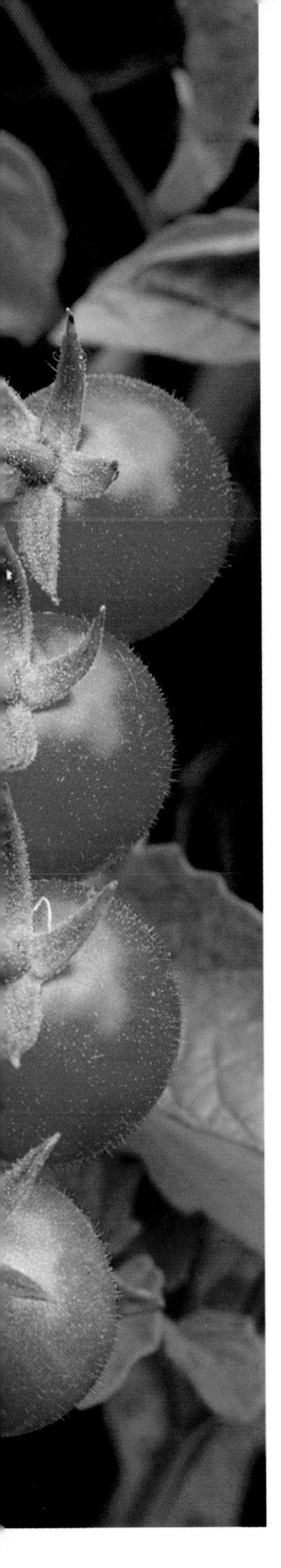

Phaseolus BEAN

From Central and South America — excavations in Mexico show that they have been grown for at least 2,000 years — runner beans (*Phaseolus coccineus*) were in European gardens probably early in the sixteenth century, though Tradescant the Elder is sometimes credited with their introduction. For a century or two, they were grown only for their brilliant red flowers, twined up arbors and over walls. They first appeared on the table in the mid-eighteenth century, when a botanist named Philip Miller discovered how delicious they were. Old sorts include the 'Painted Lady' of 1633. I grow it up bamboo tepees amongst the cabbages.

French or kidney beans (*Phaseolus vulgaris*) are also ancient. Radio carbon dates of 8000 B.C. have been obtained at archaeological sites in Peru, and they were certainly cultivated in North America by 5000 B.C. They reached Europe in the early sixteenth century. They were most popular in Italy, hence the early English and French names of 'Roman' beans. There do not seem to have been many varieties available, though by the early eighteenth century there were climbers as well as a few bush types. That soon changed, and by the 1880s vast numbers had come into commerce. Spectacular pod types can be found, some nearly 2ft (60cm) in length, others colored violet, yellow, or scarlet, plain or streaked. There are also many variations in seed design and decoration; variations in flavor do exist, but are less marked.

Pisum PEAS

The sorts of peas now used for drying have been found in excavations of 7000 B.C., making them as old as wheat and barley. They were in western Europe before the Roman expansion, though the Romans themselves had no clearly defined sorts.

Sugar-snap peas were grown in sixteenth-century gardens, and by the eighteenth century there were huge numbers of now-lost varieties of podded peas — lovely-sounding ones, too, with names like rose, rouncival, sickle, tufted, and Hotspur peas. In that century, much development had taken place in sweet-seeded varieties that are most suitable for eating fresh: 'Lincoln' is from 1849 and the American 'Tall Telephone' from 1878. Some others maintained by gardeners are far more ancient but not in commerce.

Spinacia SPINACH

A native plant of southwestern Asia, spinach was introduced to Britain, possibly from Spain, in the sixteenth century. It seems not to have been used by the Greeks or Romans, but was known to medieval Arab physicians. It was also used in fireworks (paper soaked in spinach juice acts as touch paper). Savoy-leafed, prickly and smooth-seeded sorts are all antique. Try 'Bloomsdale Long Standing' for the first (though it is only just an antique for American gardens) and 'Giant Thick Leaved Prickly' for the second.

Right: Formal rows of Welsh onions — whose narrow bulbs are as delicious as shallots and whose foliage is good in salads — contrast with neat rows of beans and the gorgeous untidiness of brassicas and fennels going to seed.

Opposite: The dense and colorful flowers of chives can look dramatic, especially if you use chives in the eighteenth-century manner, to edge paths or beds. If that gives you too much of one crop, edge other beds with wild strawberries, parsley, or thyme.

KITCHEN HERBS

Though the idea of a herb garden is immensely attractive, in the real garden most of the useful culinary herbs, especially the mints, tarragon, oreganos, and thymes, are not especially exciting plants to look at, and the mints, in particular, can colonize the ground in ways that make them a menace. Give the herb garden a strong form, using broad pathways (most herbs sprawl) and formal elements, perhaps clipped rosemary or bay. Antique gardens used some of the smaller and more polite herbs to edge paths and beds. Chives, parsley, and some of the tougher thymes can look excellent as edging, though unless you have an unusually large household, they will produce a far larger crop than you can possibly use.

Very aromatic herbs, such as bay, rosemary, lavender, and half-herbs like lemon verbena should, anyway, be grown near wherever you are going to sit in the yard. Some gardeners also choose to plant the smaller thymes among the paving stones of that area, but beware of them if you like going barefoot on a summer's day; thyme flowers attract bees in quantity.

The mints, of which there are huge numbers, though only one or two that are necessary for the kitchen, are best grown in large containers — in my garden ginger mint makes spectacular pots. Tarragon is best grown in some out-of-the-way spot (where you can put the horseradish, too) where its untidy growth will not be a nuisance. Most of the others are far better simply integrated into the vegetable garden, though the marjorams can also be planted among the flowers in the cottage border. If your garden is cool, basil is best grown indoors in a pot, where it can look (and smell) extremely good.

Allium CHIVES

Native from Britain to China and even North America, chives were probably gathered from the wild until the sixteenth century. By the eighteenth, they were used as a crisp edging for beds in the kitchen garden, where they look excellent, especially when their tiny globes of pale mauve flowers are in full show. There are no named varieties, though there is a rare, and probably antique, form with white flowers. Take the shears to them when the flowers fade or you will have huge numbers of seedlings.

Above: Dill, an annual, looks like a small fennel but has a very different and distinctive flavor, essential in many dishes, especially ones from the northern European cuisine.

Anethum DILL

Dill was much used all over Europe in renaissance medicines, especially for curing flatulence, and various female ailments and also as an aphrodisiac, though too much of it was thought to impair the eyesight. Dill is essential in every kitchen garden, used particularly in marinades for salmon and other grand fish, to flavor pickles, and for sauces. It can also be found as an ingredient in a surprising number of perfumes.

Coriandrum CILANTRO, CORIANDER

Popular in the kitchens of ancient Rome, cilantro may have been domesticated from wild plants in Asia Minor. Seed was purchased for Hampton Court in 1567, and the herb was grown on a field scale in Norfolk in the eighteenth century. Dried crystallized seedpods were used to perfume the breath and to aid digestion. In the kitchen, seeds were added to the basting juices for roast meats. Grow a row or two among the vegetables.

Mentha MINT

All the mints have been widely used since ancient times, probably more for real or supposed medicinal properties than in the kitchen. Pennyroyal was the most admired, being used to dampen hysteria, speed the menopause, and disperse fleas and flies. Spearmint, used for sauces on lamb and as a flavoring for pea soup since earliest times, was thought the most useful by Pliny. He said that it was good for the memory, discouraged worms, and stopped milk from curdling in the hot Roman summer. The Romans also rubbed peppermint on their tables before a meal, perhaps to ensure good digestion, as well as to preserve a fresh smell.

Ocimum BASIL

Originating in tropical Asia and Africa, basil is a plant of great mystery. Its full name of "basilikum" means king. It may have been associated with kingship, for there were taboos surrounding it, evidenced by its association with snakes, worms, scorpions, maggots, headaches, and blindness. Perhaps it was thought safe for only kings to handle. Even by 200 B.C., the ancient Greeks thought it hurtful for stomach, eyes, and wits. Other ancient beliefs suggested that if powdered basil was put under a stone, it bred serpents. Seed, it was said, should be sown with curses.

Left: The pinkish purple young flowers of cilantro look good if the plants are sown in rows between rows of 'January King' cabbage, or between red-leafed cabbage or leeks.

Below left: Some of the mints — this one the variegated apple mint, admired by connoisseurs of both kitchen and garden — are decorative. Try ginger mint, too.

Below center: *Mentha spicata*, like all its relatives, spreads wildly, so grow it in tubs or old buckets sunk to ground level. To dry it, harvest the stems before flowering.

Below right: Seedling basil plants are used at this size by Italian cooks for pesto sauce. It is worth sowing a pinch of seed regularly through the summer.

Whatever its disadvantages, by the eighteenth century it was widely used in salads and ragouts. The herb includes types with purple, frilled, or savoyed leaves. The purple 'Red Rubin' is similar to ancient forms, and there is also a red form of holy basil (*Ocimum tenuiflorum*). Grow the ancient bush basil (*Ocimum minimum*) and you will have neat domes of greenery that flower as summer ends. The flavor is as good as that of the more familiar culinary varieties.

Origanum MARJORAM, OREGANO

Marjoram is often said to be a mid-sixteenth century introduction to the garden, but by then it had such huge numbers of medicinal as well as culinary uses that it must be very much older. Its oil has been used to stop toothache, loosen stiff joints, and perfume the hair, and it has also been used as a face powder, for snuff, and in ragouts. The flowering tops of the plant, which have a more interesting savor than the leafy parts, were once used to dye woollen cloth purple and linens a reddish brown. In the eighteenth century, many garden variants were grown, with crisped, frilled, or variegated leaves. Of the latter, 'Gold Tipped' is the most attractive. Many are vigorous and need clipping to keep them in order. Use different sorts to create pretty knot gardens or leave clumps to flower and attract every bee for miles.

Petroselinum PARSLEY

Parsley is the most widely grown herb and has been since at least the end of the seventeenth century. It must have been introduced by the Romans, who used it as a flavoring and in medicines. They grew both flat and curly kinds (the flat sort is still widely used in continental Europe and has a slightly lighter and sharper flavor). Perhaps because the flat sort was more easily mistaken for one of the poisonous umbellifers, it was used at funerals — Pliny said that the Sardinian parsley was poisonous. Even in the sixteenth century, parsley was sometimes accused of causing epilepsy. Stalks of flat-leafed parsley, sometimes sold as "French parsley," can be blanched like celery to make a vegetable, and it has a much better taste, green in salads, than curled-leaf parsley. "Hamburg parsley" is a variant grown for the swollen root, which was eaten as winter fare by the late seventeenth century.

Thymus THYME

This herb comprises several species and many hybrids, all part of a huge and complex genus. Most of the highly aromatic species are native to Mediterranean countries. In ancient Greece, thyme was thought to give courage and revive the spirits. Thyme honey was then, and still is, much admired. The Romans made extensive use of thyme as a medicine, much of it surviving into the nineteenth century, including for melancholia, epilepsy, hangovers, toothache, and insomnia.

In the kitchen, it was also used as a flavoring for roasts and stews and even to flavor sheep on the hoof; thyme-fed lamb was an expensive delicacy. Renaissance gardeners liked to plant borders of thyme around kitchen-garden paths, while their eighteenth-century counterparts bordered flower beds with both gold- and silver-leafed varieties (*Thymus* x *citriodorus* 'Aurea' and *T.* x *citriodorus* 'Argentea,' respectively).

Opposite: Marjoram, essential for the kitchen, is easy to grow from seed and is perennial thereafter. A vigorous subshrub, its late summer flowers attract bees and butterflies.

Above left: Thymes come in a wide range of shapes, sizes, and savors. This is one of the large-leafed culinary sorts.

Above right: Curly-leafed parsley remains a good ornamental and medicinal plant, even though many gastronomes now prefer the flat-leafed variety.

A Productive Potager

Above: Apples trained on arches and a formal layout based on a design of 1618 make for a perfect back yard.

Right: A glass and iron cloche for forcing lettuces and young vegetables makes a pretty addition to the scene. The ones here date from between 1820 and 1860.

This village kitchen garden not only contains many ancient varieties of vegetables, herbs, and fruits,but also has a layout based on an antique design. One of the woodcuts in William Lawson's *A New Orchard and Garden*, written in 1618, shows a plan for just such a walled garden, with diagonal cross walks and a central bed. Lawson is so eloquent on his subject that readers will find it hard to plant anything as unproductive as a flower.

Modern gardeners might not care for Lawson's suggestion of statuary or consorts of viols among their cabbages, but we can surely not disagree with him when he writes, "Whereas every other pleasure commonly fills some one of our senses, and that only with delight; this [the kitchen garden] makes all our senses swim in pleasure, and that with infinite variety... View now with delight the works of your own hands, your fruit trees of all sorts, loaden with sweet blossoms, and fruits of all tastes, operations, and colours: your trees standing in comely order, which way soever you look... your borders on every side hanging, and dropping with Feberries, Raspberries, Barberries, Currans, and the roots of your trees powdered with Strawberries, Red, White, and Green, what a pleasure this is!" Indeed.

The bones of this ornamental kitchen garden were laid down in 1987, and though it looks large, it is only about 30ft (9m) square. It has the benefit of a sunny, south-facing site, which is more or less level (though kitchen gardens like this can be adapted for terraced plots); it also has a little shelter from other parts of the garden and from adjoining houses, though even with that the owners have found that standard-pruned currants have been beheaded by the wind. The soil is a good heavy loam.

Above: Modern ruby chard looks no different than antique sorts (*see p. 113*), so it is perfectly acceptable in an antique kitchen garden.

Left: Lettuces and leeks are allowed to run to seed to show the additional beauty of kitchen-garden crops.

The garden is given form by the outer framework of boxwood hedging, allowed to grow 1½ft (45cm) high and deep. The paths of concrete slabs are edged with thick growths of Welsh onions and chives. The circular central bed, devoted to herbs, is edged with rope-molded Victorian garden tiles. Each of the entrances into the garden has been given a handsome and sturdy metal arch, large enough and deep enough to support cordons of old varieties of apple and pear.

Each arch has four cordons of fruit on each side, allowing a good range of old varieties to be grown in a relatively small space. Among the delights featured here are, of apples, 'American Mother,' 'Devonshire Quarrenden,' 'Ellison's Orange,' and 'St. Edmund's Russet'; and of pears 'Beure Superfin,' 'Glow Red William,' and the tiny but delicious 'Jargonelle.' Though the winter and summer pruning is time-consuming, the pleasure of having that range of smells, colors, and tastes is, as Lawson wrote, a considerable compensation.

Alpine strawberries are used to cover the ground beneath the trees, something else that William Lawson suggested. For more height within the garden, wirework frames and obelisks are used to support runner and climbing beans. As the vegetation of these is cleared away after the first frosts, the obelisks give winter structure of their own. The other garden props help in winter, too, though many are also functional. The modern seakale pots are used for blanching the emerging shoots of seakale, a delicious and perennial

Opposite above: A globe artichoke in full flower; harvest it before the flower opens if you want to eat the buds.

Opposite below: Evening light on the potager. Attractive modern copies of the cloches can easily be found.

relative of the cabbage, popular since the eighteenth century. Except where there are perennial vegetables such as artichokes, seakale, and asparagus, the vegetable beds are redesigned and replanted every season. Perennial vegetables are left in permanent rows. As the garden is planned to yield salads, vegetables, and herbs as early and late in the season as possible, the pretty Victorian cloches, forcing pots, and bell jars are in active use. The garden produces more than the owners need, and the unused lettuces, chicories, and beets are left to show the potential of their flowers and seedheads. To stop the boxwood hedges from taking too much nourishment from the soil, they are root-pruned every season or so with a spade which slices down along the length of the hedges.

All potagers are helped by a little color from flowers as well as from vegetable and salad foliage. Here, the flowers are all edible ones. Borage and marigold are both antiques, though this garden uses the late-Victorian variegated form of the ancient nasturtium. The unusual herbs and vegetables include the variegated land-cress (*Barbarea verna* 'Variegata'), popular in Victorian bedding schemes, though it also makes good eating; 'Ragged Jack' kale; the delightful pink-and-white runner bean 'Painted Lady'; the blue-podded pea; and a fascinating and rare red-leafed sorrel I've not come across before, with large spear-shaped leaves in a fine mahogany red. This is certainly a plant with a past and even more certainly one with a future.

Above left and center: Boxwood hedges frame the design, and wild strawberries (*Fragaria vesca*) edge the paths. The center of the garden is marked by a grand terra-cotta pot, which will eventually contain a topiary bay tree.

Above right: The wild European strawberry produces runners, though these are easily controlled.

THE CONTAINER GARDEN

Plants have been grown in containers for at least three thousand years, and today almost every gardener makes use of at least a few plants in pots. In ancient China of 1000 B.C., pot plants were domestically important, expensive, and commonly given elegant containers. In the courtyard houses of ancient Greece, pot plants were used for decoration, though they were also functional: at the Hephaisteion in the Agora of Athens, a row of unidentified trees was planted in large pots and sunk in cuts in the rock to make an avenue.

Roman horticulture, too, made extensive use of containers, not only to grow citrus fruits and vines, but also to help with forcing cherries and figs (slaves had to move the pots around to keep them in the sun against a warm wall). Potted fruit trees were carried into the dining space when the fruit was ripe. Wall paintings show that decorative containers were also used for roses, pomegranates, and oleanders, and we know that plants were transported around the empire in pots.

Opposite: Lilies, boxwood topiary, Paris daisies, and tiny oranges are all part of this assemblage of containers in a city garden. Above: Potted auriculas in an enchanting painting of 1817, by Pancrace Bessa.

In the Middle East at the same time, garden pots were used in the Roman manner for planting out such things as lemons and grapevines, and numbers have been found at the winter palaces of Herod Tulu Abu el-Alayik, south of Jericho. Farther west, similar pots have also been discovered at the Roman excavations at Watling Street in London.

Grand medieval gardeners continued the Roman use of pots (large numbers of earthenware pots were ordered for Hampton Court in 1515); poorer ones used containers of wattle or crude earthenware and had them on windowsills, by the well, or (sometimes planted with fragrant carnations or pinks) next to turf seats. Once excavations of ancient Rome began during the Renaissance, Italian grandees copied Roman models (often funerary urns) and grew elegant potfuls of pomegranates, lemons, or bitter oranges to decorate their loggias, staircases, and grand entrance halls.

In prosperous late-medieval gardens, at least in southern Europe, there was often a "winter house" where valuable potted plants were kept during that season. Such areas had a fireplace, so lemons and oranges managed to survive both cold and dark until late spring. Most citruses coped well, even when the only thermometer was a saucer filled with water — when it froze, fires were lit. Throughout Europe, big collections of varieties were kept and maintained as being appropriate to the gardens of the wealthy. By the thirteenth century, the Chiaramontesi garden in northern Italy had several thousand lemons and oranges. In Renaissance mansions, conditions for overwintering oleanders, myrtles, pomegranates, and brand-new American exotics such as agaves and opuntias were at least slightly better, for the handsome and well-lit pavilions used for the owners' summer banquets were often transformed into the winter quarters for exotic "greens."

Once glass became cheaper, a century or so later, it could be used for almost any garden building: summerhouses, plant houses, and primitive orangeries, which had just begun to appear in the late sixteenth century, became common throughout Europe in the 1600s. Container gardening soon became an important skill. In a few decades, tender plants from all over the world were grown, often outdoors during the summer and overwintered under glass.

By the beginning of the nineteenth century, cast-iron technology (part of the foundation of the Industrial Revolution) had improved so much that log-burning stoves, heating pipes, and even the greenhouses themselves could be manufactured easily and cheaply. Soon, the rich had plant houses that could duplicate the most tropic climes and that they could fill with orchids, ferns, palms, and other exotics from anywhere in the world. Only a few decades later, even middle-class houses had conservatories in which the owners could cultivate rare plants bought in the imported plant auctions that took place in almost every major city in Europe. The Industrial Revolution also made plant containers cheaper, and soon every house and garden in Europe and North America had a few geraniums or ivies in a tin, cast-iron, or terra-cotta urn.

Above: The delicious and early-fruiting 'May Duke' cherry.

Opposite: The interior of a Victorian vine house. The bed of soil at floor level would have been used for forcing early vegetables or even for growing pineapples. The grapevines were planted with their roots outside. Heating was probably provided by chimney flues running up the back wall.

Gardening with Containers

Above: Berries and variegated foliage of *Aucuba japonica*, a popular plant in nineteenth-century city gardens.

Right: Aucubas are easy to grow from cuttings if you want to copy this amusing arrangement of pots.

Good places for pots in informal gardens include by doorways, on low walls, around a sundial, or in clusters by sitting areas. In topiary gardens, especially in large ones, pots can frame a view or even mark their boundaries. In the kitchen garden, they can mark the junction of paths or the center of beds or stand beside a bean arbor. They make good homes for invasive herbs like mint or floppy ones like marjoram. The rose garden, sometimes difficult to structure and often a bit flat once the main flowering season is over, can usefully be decorated with all sorts of containers — either for topiary, to give strong architectural emphasis, or for sumptuous flowering plants to give color until the fall. Under glass, containers should be a movable feast.

In gardens for antique plants, it is worth planting in classic types of container. Terra-cotta garden pots are widely available, from tiny to huge. These pots look splendid and soon appear suitably aged. Plants grow well in them and are easily repotted except when the pots are enormous. The pots can be decorated; swags of foliage and satyrs' masks are the most traditional.

"Versailles" tubs, cube-shaped with decorative finials and handles for mobility, look equally good with topiary, tree standards (they were used at Versailles for an ancient collection of citrus trees, purchased for Louis XIV), or a tumble of agapanthus. Although they are commonly painted dark green, some old Italian gardens have the framework in gray and the panels in olive.

When choosing a container, first consider whether it is to stand outdoors all year and, if so, what your winters are like. While some terracottas can stand reasonable amounts of frost, serious cold will flake off the rims or crack them apart. Wintered indoors (even in a shed) they survive perfectly well.

Plastics are winterproof, but weather badly, becoming dirty rather than attractively patinated. Fiberglass ones do better, and the better imitation Versailles tubs and painted or metal urns make good "focal points" outdoors all year. Metal containers need some care; if the soil gets waterlogged and then freezes, cast-iron can split.

Any of the plants mentioned in this book can look handsome in a container. Experiment. Fill pots with bulbs: many lilies fare better in pots than in an exposed bed (*Lilium tigrinum* does wonderfully, though try the nineteenth-century *L. longiflorum,* too). Many annuals also do well: try big pots of *Nicotiana affinis* or *N. sylvestris*, or smaller ones of Chinese annual dianthuses.

The following are some plants not covered elsewhere in the book that thrive in containers. Some are essentially hardy in northern gardens, but can be moved under cover in harsh conditions. Others are more or less tender, depending on your climate. With tough winters, even plants like the species of *Agapanthus*, particularly forms of *A. africanus* (early nineteenth century), need to be protected, and in northern gardens, ancient pot plants like the oleanders barely have enough summer sun to make them flower. There, too, most citrus plants need to be sheltered all year, though some, like the gorgeous 'Imperial' lemon, will appreciate a month or two outdoors to produce flower buds. *Canna indica* (look for the eighteenth-century plant with variegated leaves) can cope, but northern winds batter the flowers, making shelter essential. Some fuchsia species are hardy in the north, but many antique hybrids need to be indoors for most of the year. Rhododendrons, too, have marvelous antiques (try 'Lady Alice Fitzwilliam' and 'Princess Alice') that need to be in pots, though the perfume is so good, you won't mind caring for them.

Above: The calamondin (*Citronella microcarpa*) is easy to grow in a bright, frostfree place, and its flowers smell just as good as those of lemons and grapefruit. The smallish "oranges" can be used like kumquats.

Right: The small maples, especially forms of *Acer palmatum*, look splendid in pots, especially in small, sheltered, formal spaces. All of them have brilliant foliage in the fall.

Below: Paris daisies, this one the basic species, are easily clipped into shape early in the season and then left to flower for the rest of the summer.

Acer MAPLE

Small yards can often be heavily shaded, but they are also sheltered. If you need interesting foliage and powerful fall colors, then try some of the myriad varieties of Japanese maple (*Acer palmatum*). These were immensely popular in Japan and are found painted on medieval screens, though their greatest popularity was in the seventeenth and eighteenth centuries — the first Japanese book about them was published in six volumes in 1710. Choice varieties began to appear in the West from around 1820; few remain, but try 'Osakazuki' from around 1840 or the smaller and purple-leafed 'Umegae' from a few decades later. Provided they don't dry out, they make fascinating plants for large pots or tubs.

Argyranthemum PARIS DAISY

Argyranthemum frutescens, in its simplest yellow-flowered form, was introduced to Britain from Paris in the late seventeenth century, hence its English common name, though the basic species is native to the Far East. Other variants soon arrived. The oldest sorts, with single white flowers and jagged olive-green foliage, make marvelous plants for large tubs or vast earthenware pots. Clip into domes when they become messy.

Brugmansia suaveolens DATURA

This species has marvelous pendant white flowers, sometimes almost 12in (30cm) long, heavy with perfume at night. From Central America, it was in eighteenth-century European glasshouses; the double white *Brugmansia* x *candida* may have been grown since the seventeenth century. The annual *B. meteloides*, a sprawling bush if well fed, has enormous white and perfumed flowers, making it a good substitute if you have nowhere to overwinter the other species.

Crinum CRINUM

The largest and handsomest of the antique bulbs is the dramatic Orange River lily (*Crinum bulbispermum*). Taken to Europe from South Africa probably in the seventeenth century, it makes luxuriant clumps of large leaves, from which arise stems up to 5ft (1.5m) tall, topped with large flowers, which are usually white, but sometimes striped in pinkish purple. The more recent hybrid *C.* x *powellii* is a rich candy pink. The perfume of both is delicious and heady, especially on warm summer evenings.

Thought to be tender when first introduced, crinums were often grown under glass all year, when several crops of flowers were obtained. Dry plants off in winter.

Dendranthema CHRYSANTHEMUM

Though there are more than thirty Asian species, the ornamental chrysanthemums (complicated hybrids involving *Chrysanthemum indicum* and others) probably originated in China long before 500 B.C. They were exported to Japan about 800 A.D. and to Holland in the late seventeenth century. The first ones to arrive in Europe were lost, but others reached Britain in 1790, causing great excitement. Between 1816 and 1823, seventeen new sorts were added, and experiments in crossing soon began. Growers produced some incredible plants: one, in a 16in (40cm) pot, won all the prizes; it was said to be over 27ft (8m) in circumference and carried over a thousand flowers. The Chinese would have been horrified to hear this; they allowed their plants no more than four or five perfect blooms.

Few old cultivars have survived intact with their names. 'Emperor of China' is one and may be one of the ancient Chinese garden plants. But antique groups, rather than original varieties, can be identified. Old Chinese forms had masses of incurving petals, to give a globe- or a bun-shaped bloom. Old Japanese sorts had recurved petals, giving a shaggier bloom. Of European developments, there were no pompon types before 1846, no anemone-types before 1856, and no varieties with heavily reflexed petals before 1862.

Above: Pink and powerfully fragrant, this is *Crinum* x *powellii*, a nineteenth-century hybrid of the old seventeenth-century "sea lily."

Left: *Chrysanthemum* (now *Dendranthema*) 'Quill Elegance.' Quilled flowers like this had been developed for showing by the 1830s.

Dendrobium ORCHID

Once large numbers of gardeners in northern Europe and North America had greenhouses and solaria, the jungles of Asia and Central and South America were soon ransacked for beautiful plants to put in them. At first, though, the tropical warmth and humidity necessary to grow orchids were the preserve of the very rich. By 1839, the magnificent glasshouses at Chatsworth House in Derbyshire, England, were packed with orchids, many of them collected specifically for the then Duke of Devonshire. In that year, a plate was published of His Grace's *Dendrobium paxtonii*, the most magnificent orchid in cultivation. It had only just flowered, the immense plant having been shipped whole from the Amazon. It can be found in some garden centers.

By 1852, there were auctions in London of Guatemalan orchids, and other genera of orchid such as *Vanda* and *Cymbidium* were becoming more widely grown.

The most gorgeous of the vandas (*Vanda rothschildianum*) is an antique, and although it is a delicate mauve, other vandas come close to that rarest of garden colors, a genuine blue, and some have purple or mottled foliage.

Some *Dendrobium* species, including *D. nobile*, were grown by the 1820s. Many are easy to grow and reflower. In northern Europe, dendrobiums and some cymbidiums can be grown outdoors in shade during the summer and brought indoors in September. Many will flower soon thereafter.

Although some orchids need surprisingly dry conditions, many still do best if you can re-create the lush surroundings of their native tropics. These varieties will benefit from being mist-sprayed once a day, but the soil should never be allowed to get sodden.

Opposite above and below: Gorgeous hybrid cymbidiums like these were soon developed from species being imported to Europe in the early nineteenth century.

Left: The sumptuous *Vanda rothschildiana* was named in the mid-1800s after that orchid-loving dynasty.

Above: Seedlings of the 'Marine' strain of heliotrope can give deep purple plants. One this size should perfume a small patio.

Heliotropium HELIOTROPE

Though they do well outside during the summer months, heliotropes can be kept in flower until early winter indoors, where the smell is even more delicious. They reached their greatest perfection and variety in Victorian gardens in summer and in their sunrooms in winter. Heliotropes are essential for any garden of the period.

Most varieties are hybrids between species of *Heliotropium* that arrived from Peru in the 1750s. They made little headway until the second decade of the nineteenth century, when Regency gardeners had good orangeries and glasshouses. Modern named forms, which range from deep purple through palest mauve to white, are all comparable to their nineteenth-century ancestors and may be identical.

Above left: *Pelargonium* 'Miss Burdett Coutts' shows how colorful even the leaves of some nineteenth-century varieties can be; the flower buds were often pinched out.

Above right: Pungently lemon-scented *Pelargonium crispum minor* grows luxuriantly outdoors in a big pot, but does just as well in a small one on the windowsill.

Hydrangea HYDRANGEA

This is a large genus of shrubs and a few climbers, with species growing in Asia and North America. The American *Hydrangea arborescens* makes a wonderful tub plant at Hidcote Manor in Gloucestershire, England, and in most yards will have flower trusses in good condition until the late fall. It was introduced to Europe in 1736. There were many more arrivals, mostly from China and Japan, in the following decades. There, many varieties were ancient garden plants, but in Europe breeding began only at the end of the nineteenth century. The plant called *Hydrangea* 'Joseph Banks,' with its vast heads of greeny cream flowers, later pale pink or hyacinth blue, is probably an ancient oriental garden variety and arrived in the West in 1789. It is excellent in containers.

Pelargonium GERANIUM

"Geraniums," still so called as they were once placed in that closely related genus, have been grown since *Pelargonium triste* (the "sad-colored" pelargonium) arrived in Europe in 1631. Like most species it comes from South Africa. Unusually, it is tuberous-rooted, with sprays of delicate yellow and brown flowers. It is not especially showy but worth growing for its powerfully sweet scent of cloves and other spices, strongest at night.

By the early years of the nineteenth century, the geranium had become "the chief of flowering plants suitable for the parterre." Many nineteenth-century varieties still exist, including 'Happy Thought' (green-margined leaves with a gold heart and brilliant scarlet flowers), 'Freak of Nature' (a dwarf variety), 'Crystal Palace Gem,' 'Golden Harry

Hieover' (fine yellow leaves with a grayish horseshoe, often used for edgings in the parterre), 'Mrs. Parker,' 'Caroline Schmidt' (with almost silver leaves), 'Lass o' Gowrie,' 'Miss Burdett Coutts,' 'Mangle's Variegated' (probably the most famous of them all, a dwarf variety with dark green, silver-splashed leaves), and 'Mrs. Pollock' (yellow foliage with a horseshoe of reddish bronze).

Ivy-leaf geraniums also began to arrive in the early 1700s, and some of the cultivars have been long-lasting; 'L'Elégante,' with silver-edged and variegated foliage and pale violet flowers, was praised in 1887 and is still around.

Regal geraniums, called Martha Washingtons in the U.S., seem to have originated at the royal residence Sandringham House and reached less grand windowsills around 1877.

Scented geraniums began to arrive in Europe in the seventeenth century, but most appeared in the next. *Pelargonium* 'Fragrans,' a hybrid with pine-scented gray-green leaves, has been known since 1645. The common rose-scented *P. graveolens* was introduced in 1774, as was the even rosier *P. radula.* Peppermint- and lemon-scented types have been around since the late eighteenth century, the variegated 'Lady Plymouth' since 1802, and the deliciously orange-peel-scented 'Prince of Orange' since at least 1880.

Verbena VERBENA

The half-hardy perennial bedding verbena (*Verbena* x *hybrida*) makes marvelous potfuls indoors, many pleasantly perfumed and all colorful. They are the result of crosses between species introduced from South America between 1826 and 1837. Hybrids, popular by 1844, include the scented and soft pink 'Silver Anne,' the deep purple 'Hidcote,' and the loud scarlet 'Huntsman' — all modern names for plants available from the 1850s. Many look good combined with white Paris daisies and perhaps gray-foliaged helichrysums. Indoors, aphids can be a nuisance.

Below left: *Hydrangea paniculata* and most other shrubby hydrangeas grow well and look good in large tubs or pots.

Below right: These striking verbenas — the purple 'Huntsman' and scarlet 'Hidcote' — make a startling combination in a medium to large pot.

A Victorian Conservatory

The house and garden, in the English Cotswolds hills, are old, but this conservatory is a mere stripling of 12 years. It replaces a handsome vine house of 1848 that had (as conservatories do) reached the end of its days. The new structure follows the plan of the earlier one and even incorporates its fireplace (though the owner complains that it smokes if the wind is from the wrong direction). The room is divided by a partition into two sections.

Below: The entrancing antique Wardian case — a conservatory within a conservatory.

Above: The scented flowers of *Passiflora caerulea*, an easily grown antique climber.

Below: Many ferns make excellent conservatory plants for shady corners; they were Victorian favorites.

One, barely heated, is used for the owner's collection of prize-winning show auriculas; the conservatory doors are left open all summer to prevent overheating. The other section is heated and allows the luxuriant growth of exotic climbers and pot plants. It is filled with containers growing mostly antique plants, as well as some well-chosen antiques. It makes a perfect place to sit on a summer's morning — or a winter's evening, with the fire lit and a few candles.

The conservatory's most important element is light. Being south-facing, it is potentially as hot as Death Valley in summer, but instead of blinds to create shade, here it is the plants themselves that perform the function, producing a result much more sympathetic to human inhabitants and to the plants growing beneath. However, plant maintenance is more work than looking after blinds.

In this conservatory, the climbers run over a free-standing metal arbor that has been specially designed to support them, but it is easy to make something that gives a similar effect. Simply provide the roof beams with wires along their length and support the wires from brackets so that they are at least 6in (15cm) from the glass; this gives the leaves plenty of room without their

Above left: The wall fountain belonged to the original conservatory; the weather vane is part of the owner's collection. The pale-flowered orchid is *Dendrobium paxtonii*.

Above right: A stone shepherd, one of the many decorative antiques that adorn this indoor garden, stands shaded by grapevines.

Above: The headily perfumed and easily grown double datura (*Brugmansia* x *candida* 'Plena') will flower well into early winter.

Right: The division between the barely heated and the well-heated areas in this dual-purpose conservatory. The Wardian case provides another layer of insulation for the plants it contains.

touching the glass itself. The climbers can then be trained up the wires, with much of their side growth kept pruned back.

Most of the climbers grown here are grapevines. One is eight years old; the other is a shoot that springs from the remaining roots of one planted by the original conservatory. Grapes need a substantial amount of work through the season if they are to remain pestfree and to give a good crop. If you prefer less effort, use a pest-resistant plant, like a perfumed jasmine (*Jasminum sambac* or *J. officinale*). All require thinning after each wave of flowers, but do not need spraying. Alternatively — and especially if you want to use the conservatory during the winter and need more light — grow vigorous climbers that are often treated as annuals: morning glories or the marvelously scented moonflower (*Ipomoea alba*).

Another good climber in this conservatory is *Plumbago capensis.* Easy and fast-growing, it often puts on 5–6½ft (1.5–2m) growth in a season. The china blue (or, more rarely, white) flowers last several weeks, though bunches are produced through summer and fall. They have, alas, no perfume. For that, try *Hoya carnosa*, which will hardly flower when it is growing fast in a large pot but blooms when it is starved. The pinkish, waxen flowers have an overwhelming smell in the evening. If that is too much, try the gentler but glorious scent of the passionflowers. The one here is the lovely antique *Passiflora caerulea*, which is almost hardy, with pale greenish-blue flowers.

Once green summer shade is provided for, add to the generosity of greenery by using shelving or tiered stands to pack in yet more foliage, especially of shade-tolerant plants. Though some of the shelving systems in this conservatory are from the late nineteenth century, modern examples can be just as effective and are more easily found. Fill them, as here, with lots of pots of streptocarpus. Easy from seed or cuttings, the plants thrive in these conditions and have long flowering seasons, dramatic flowers, and large, splendidly veined leaves. They seem to enjoy close contact and do well crammed onto shelves and tables.

An amusing contrast here is provided by the little bead plant (*Nertera granadensis*), with its tiny leaves and large red berries. Popular in Victorian garden rooms, it is easily grown from seed and is sometimes found in garden centers. It does best if watered from below; the leaves and stems rot if watered from above.

The effect of any container garden, whether outdoors or under glass, is partly determined by the props: seating, statuary, types of container, and so on. Here, decorative antiques (the owner deals in antiquities) combine delightfully with the plants. Choose things that won't be damaged by high humidity or the occasional misdirection of a hose or plant mister.

The most exciting piece is, of course, the Wardian case, providing a delightful conservatory within a conservatory. Wardian cases first appeared in the late 1820s, the invention of one Nathaniel Ward, and were used for growing plants that hated conditions in the average living room. They then became immensely important as a means of transporting jungle plants around the world and were instrumental in the introduction of many plants to European and North American conservatories, especially ferns and orchids. This one is an heirloom, having once belonged to the owner's great-uncle, who used it to house part of a remarkable orchid collection.

Above: Cymbidiums like this were being hybridized from the 1850s; this one is called 'Rosanette.'

Left: Trails of *Plumbago capensis* almost smother the fireplace and make a luscious canopy for the owner's favorite chair.

Cultivation notes and sources

Most of the plants listed in this book are widely available; look in local nurseries, especially ones selling cottage-garden flowers. While every gardener's definition of what these are varies, you will almost certainly find at least a few of the plants in this book. Where plants are not widely available, a specialist source and code reference has been given; see the list of addresses at the end of these cultivation notes. Prices of each company's catalog are also given.

ACANTHUS
(Bear's breeches)
Mostly deciduous, long-lived perennials, height up to 5ft (1½m), spread up to 6½ft (2m). Flower spikes appear in late summer or fall in green, purple, and white, but grown largely for their spectacular leaves. Tolerates full shade; hardy to zone 6. Any soil that is not waterlogged. Propagate either by division, by root cuttings, or by seed. Widely available.

ACER
(Maple)
Shrubs or small trees, with height and spread from 6½ft (2m) to 16ft (5m). Flowers often pinkish, leaves variable in form. Shelter and shade; some hardy to zone 3, others only to zone 7. Soil average to moist; don't allow to dry out or stand in windy places. Propagate by grafting, seeds. Widely available.

ACONITUM
(Monkshood)
Herbaceous perennial, rarely herbaceous climber; *A. septentrionale* 3ft (1m) × 2ft (60cm), *A. napellus* 8ft (2.5m) × 1ft (30cm). Flowers in summer/late summer, in pale yellow, blues, muddy pink, white. Part or full shade; *A. septrentionale* hardy to zone 3, *A. napellus* to zone 5. Rich soil, dry to moist. Divide every few seasons. Propagate by seed; rootstocks easily divided. *A. napellus* widely available; *A. septentrionale* sold as *A. lycoctonum* ssp. *lycoctonum*. MI01, WA04.

ALCEA
(Hollyhock)
Short-lived perennial, 6½ft (2m) × 3ft (1m). Flowers in high summer, from white to black, mainly yellows and reds. Sun; hardy to zone 3. Rich soil, average moisture. Susceptible to rust. Propagate by seed, or division for anything special. All widely available.

ALLIUM CEPA
(Onion)
Bulb, up to 1½ft (45cm). Sun; most sorts hardy zones 4–9, but will rot in damp soil in winter. Good garden soil, average moisture; sow seed or plant sets in spring, lift bulbs in late summer. Widely available.

ALLIUM PORRUM
(Leek)
Leafy bulb, often grown as annual, up to 1½ft (45cm). Foliage in greens or blues, rarely purple. Sun; hardy to zone 4. Good garden soil, average moisture; sow early spring, harvest through winter. Propagate by seed, rarely side bulbs. Widely available.

ALLIUM SCHOENOPRASUM
(Chives)
Clump-forming herbaceous perennial, height and spread 1ft (30cm). Flowers in summer, in purple, rarely white. Sun or part shade; hardy to zone 3. Divide clumps every couple of seasons. Propagate either by division or by seed. Widely available.

AMARANTHUS
(Prince's feather)
Colorful leafy annual, up to 2ft (60cm). Foliage in yellows, greens, scarlets. Likes sun; tender. Rich garden soil, average to moist; sow seed in spring, water to avoid early bolting. Propagate by seed. Widely available.

ANEMONE
Spreading perennials, up to 5ft (1.5m) in *A. hupehensis*, only to 8in (20cm) in others listed. Flowers white, pinks, purples, blues. Species listed like full sun, though *A. hupehensis* will do in part shade; *A. hupehensis* hardy to zone 5. Good garden soil, average moisture; divide clumps of *A. hupehensis* every few seasons. Propagate by division. All widely available.

ANETHUM
(Dill)
Annual, to 2ft (60cm) × 1ft (30cm). Sun or part shade; tender. Good garden soil, average moisture. Propagate by seed. Widely available.

ANTIRRHINUM
(Snapdragon)
Short-lived perennials, often treated as annuals, up to 3ft (1m) × 2½ft (75cm). Flowers in summer/late summer, in white, pinks, reds, yellows. Sun; will survive light frost. Rich garden soil. Propagate by seed, though take cuttings of favorite varieties. Widely available.

AQUILEGIA
(Columbine)
Short-lived clump-forming perennials, up to 3ft (1m) × 2ft (60cm). Flowers in late spring and early summer, all colors and many combinations. Sun or part shade; hardy to zone 4 or 5. Keep named sorts deadheaded. Propagate by seed. Widely available.

ARGYRANTHEMUM
(Marguerite daisy)
Shrubby perennials, 3ft (1m) × same. Flowers in summer and fall, all winter indoors, in pinks, white, apricots, yellows. Needs sun; hardy to zone 9. Needs rich soil. Feed during growth. Propagate by cuttings. Widely available.

BELLIS
(English daisy)
Low, spreading perennials, up to 6in (15cm) × indefinite. Flowers spring to fall, in white to red; one form has variegated leaves. Sun or shade; *B. perennis* hardy to zone 4. Divide regularly. Widely available.

BRASSICA
(Broccoli, cabbage, cauliflower, etc.)
Leafy biennial, usually harvested in first season. Height varies, but up to 3ft (1m) × same. Flowers important only in broccolis and cauliflowers, foliage in greens, blues, purples. Sun; some varieties hardy, others not. All varieties like good garden soil, average moisture; sow seed or plant seedlings in spring. Many varieties widely available.

BRUGMANSIA SUAVEOLENS
(Angel's trumpet)
Can become a small shrub, but will flower in a 12in (30cm) pot, up to 16ft (5m) × 10ft (3m). Flowers outdoors in midsummer until the first frosts; under glass into early winter, in white, pinks, or amber yellows. Likes outdoors — full sun best; under glass, some shade; hardy to zone 10. Rich soil, feed weekly in pots, keep moist, watch for aphids and snails. Propagate by 15cm (6in) cuttings in summer and early fall. Widely available, often listed as *Datura suaveolens.*

BUXUS
(Boxwood, box)
Evergreen bushes, usually clipped, but some sorts up to a height and spread of 13ft (4m). Foliage can be variously variegated. Sun or shade; hardy to zone 10. Propagate by cuttings at any time of year. Many varieties widely available.

CALENDULA OFFICINALIS
(Calendula, pot marigold)
Annual, sometimes biennial, 1ft (30cm) × 2ft (60cm). Flowers in summer, fall, overwintering plants in spring, in shades of yellow, gold. Sun; easy in any garden soil. Widely available.

CAMASSIA
Bulb, up to 3ft (1m) × 1½ft (45cm). Flowers in early summer, in shades of blue, sometimes cream. Sun; hardy to zone 3 or 4. Easy to grow. Propagate by division of clumps or by seed. Widely available.

CARDAMINE
(Bitter cress)
Herbaceous perennial, height and spread 1ft (30cm). Flowers mid- to late spring, in various shades of mauve, single, rarely double. Sun or part shade; hardy to zone 3. Soil moist to wet. Double form needs division every season or two. Propagate by seed or division. CT01, W-MI02.

CHEIRANTHUS
(English wallflower)
Short-lived perennial, treated as annual or biennial, height and spread up to 2½ft (75cm). Flowers in mid to late spring, in tawny brown, yellows, white, reds. Likes sun; will survive moderate frost. Average to poor soil, average to dry. Propagate old doubles by taking summer cuttings. Widely available as *Erysimum cheiri* or *Cheiranthus cheiri.*

CLEMATIS
Perennial climbers and scramblers, to 35ft (10m) × indeterminate. Flowering depends on species and variety, late winter to mid-fall, in white, yellow, red, violet. Most like partial shade; hardiness varies, but most are hardy to zone 4. Normal soil, average moisture. Many

like root area shaded by other plants. Propagate by layering, seed, or cuttings. Widely available.

COLCHICUM

(Colchicum, naked boys)

Fall-flowering bulbs, foliage in spring and summer, biggest sorts 1ft (30cm) high. Flowers throughout fall in white, purple, amethyst, sometimes double. Hardy to zone 4. Divide every three seasons in late winter. Propagate by division, or seeds. Widely available.

CONVALLARIA

(Lily-of-the-valley)

Creeping perennial, height 6in (15cm), indefinite spread. Flowers in spring, white to red; leaves can be variegated. Sun or shade; hardy to zone 4. Propagate by division. Widely available.

CORIANDRUM

(Coriander, cilantro)

Annual, 2ft (60cm) × 1ft (30cm). Flowers in summer, white or pink, but grown mostly for its aromatic leaves. Sun or part shade; tender. Propagate by seed. Widely available.

CRINUM

(Crinum lily, naked ladies)

Large bulbs, 3ft (1m) × same. Flowers in summer, white, sometimes pink. Sun or part shade; hardy to zone 9. Average to poor soil, average moisture; flowers best when pot-bound. Propagate by seed, division. Widely available.

CROCUS

Bulbs, up to 6in (15cm). Flowers in spring or fall, in yellow, mauve, blue, white. Sun or part shade; most hardy to zone 3. Well-drained soil, average moisture; divide every few seasons. Propagate by division. Widely available.

CUCURBITA

(Gourds, squash, pumpkin)

Trailing or climbing annuals, indeterminate height and spread. Flowers in summer, fall; fruits in many shapes, sizes, colors. Sun; tender. Rich garden soil, average to moist, feed throughout season, harvest fruit before frost. Propagate by seed. Widely available.

CYCLAMEN

Tuberous perennials, height and spread 6in (15cm). Flowers in fall or spring, in rose pink, pale purples, white. Sun or shade; some hardy to zone 5, most only to zone 9. Prefers well-drained soil, dry to average moisture. Propagate by seed; sow outdoors in fall. Widely available.

CYNARA

(Artichoke, cardoon)

Herbaceous perennial, often grown as an annual, in flower to 2m (6½ft), foliage as much across. Harvest flower buds from summer to autumn. Sun; hardy to zone 9. Good garden soil, average moisture; in cold gardens, protect roots with straw; divide big clumps. Propagate by seed or treat side rosettes as cuttings. Widely available.

DENDRANTHEMA

(Garden mum)

Shrubby, often treated as herbaceous perennials, height and spread 4½ft (1.5m). Flowers in late summer, fall, in many colors. Sun or part shade; hardy to zone 4. Take cuttings in spring or propagate by seed. Widely available.

DENDROBIUM

(Orchid)

Epiphytes, or ground-living, from cool mountainside to tropic jungles, few more than 3ft (1m) × indeterminate. Flowering depends on species and variety, yearround, in flowers in all colors. Some sorts with purplish or mottled foliage. Shade or partial shade; Hardy to zone 10.

Epiphytes need orchid medium; terrestrials can be grown in ordinary potting mix. Some need dry conditions, none will tolerate sodden soil for long.

Dendrobiums like "intermediate" temperatures of 55–80°F (13–27°C), though can tolerate cooler winters. Shade in summer, mist daily. Best grown on bark.

Cymbidiums like "cool" conditions of 50–75°F (10–24°C) and will often grow well in ordinary rooms. To get them to flower, grow outdoors in shade during the summer.

Vanda will do best at 55–80°F (13-27°C), though can tolerate cooler winters. Shade in summer, mist daily. Best grown on bark. Propagation: clumps of pseudobulbs can be divided. Vandas can have their stems cut into sections. Widely available.

DIANTHUS

(Carnation, pink, sweet William)

Mat-forming perennials, up to 1½ft (45cm) × indefinite. Flowers in early summer, in white, pink, deep maroons. Sun; most hardy to zone 4. Does best with low nutrient levels, dry to average soil. Propagate by cuttings or layering in spring. Many widely available.

EPIMEDIUM

(Epimedium, barrenwort, bishop's hat)

Creeping perennial, evergreen, to 1ft (30cm) × indefinite. Flowers in spring, fall, and winter; leaf color bronze to scarlet, flowers yellow to rosy orange or pink. Shade; hardy to zone 5. Propagate by division. Widely available.

ERANTHIS

(Winter aconite)

Tuberous herbaceous perennial, height and spread 6in (15cm). Flowers in late winter, early spring, in yellow. Sun or shade; hardy to zone 4. Propagate by division or seed (self-sows well). Widely available.

ERYTHRONIUM

(Dog's-tooth violet, trout lily)

Herbaceous perennials, height and spread from 6in (15cm) in *E. dens-canis*, to 1ft (30cm) in *E. americanum*. Flowers in spring, in rose pink, yellows, white. Partial or full shade; hardy to zone 3. Peaty soil or leafmold, average to moist, not boggy. Propagate by seed; sow outdoors in fall. Widely available.

EUONYMUS

(Euonymus, spindle tree)

Deciduous or evergreen shrubs or small trees, to 9–10ft (3m) × 6ft (2m). Flowers in spring; grown for fall fruits and foliage. Partial shade; hardy to zone 3 or 4. Grows well in alkaline soil; use several plants to guarantee good fruiting. Propagate by seed or cuttings. Widely available.

EUPHORBIA

Annuals, biennials, creeping evergreen perennials to shrubs, 1ft (30cm) × indefinite in *E. robbiae*; to 4½ft (1.5m) × same in *E. characias*. Flowers in spring to summer, in greens, yellows. Sun or shade; perennials hardy to zone 6 or 7. Propagate by division in creeping species, by seed in others. *E. robbiae* widely available as *E. amygdaloides*. *E. characias* widely available.

FAGUS

(Beech)

Deciduous tree, can make hedges to 40ft (12m) high. Foliage can be variegated or purple. Sun or part shade; hardy to zone 4. Propagate by seed, cuttings. Widely available.

FOENICULUM

(Fennel)

Mostly herbaceous perennials, up to 6½ft (2m) × 3ft (1m). Flowers in summer and fall, leaves blue-green or purple. Sun or part shade; hardy to zone 4. Good garden soil for bulb-forming varieties, ordinary for others, average to moist. Widely available.

FRITILLARIA

(Crown imperial)

Bulbs, height up to 3ft (1m) in *F. imperialis* and *F. persica*, usually half that. Flowers appear in spring, in white, yellow, orange, shades of purple, sometimes checkered. Sun or partial shade; *F. imperialis* hardy to zone 5, others to zone 3. Propagate by seed, division. Widely available.

HEDERA HELIX

(English ivy)

Self-clinging climber, evergreen, rarely a bush form, both height and spread indeterminate. Flowers in winter; large range of leaf forms and variegations. Sun or shade; hardy to zone 5. Do not attempt to grow up damp or poorly mortared walls. Propagate by cuttings. Widely available.

HELIOTROPIUM

(Common heliotrope)

Weak shrubs, 2ft (60cm) × 3ft (1m). Flowers whenever in growth, in mauves, white. Sun, part shade; hardy to zone 10. Rich soil, average to moist. Cut back when it becomes too leggy. Propagate either by cuttings or by seed. Widely available.

HELLEBORUS

Herbaceous perennials to low subshrubs, up to 2ft (60cm) × indefinite. Flowers winter/spring, in white, green, shades of pink. *H. niger* needs good soil and light; *H. foetidus* prefers shade. Hardiness varies: *H. foetidus* to zone 7; *H. niger* to zone 4. Propagate *H. niger* by division, others by seed. These and *H. orientalis* (introduced in the 1840s) widely available.

HEMEROCALLIS
(Daylily)
Herbaceous perennial, up to 4–5ft (1.25–1.5m) × indeterminate. Flowers summer, yellow to rust red. Part shade; most varieties hardy to zone 4. Divide every few seasons. Propagate by division for named sorts; otherwise by seed. Widely available.

HESPERIS
(Sweet rocket)
Biennial, up to 4½ft (1.5m) × 2½ft (75cm). Flowers summer, pure white to pale purple. Sun or part shade; hardy to zone 4. Propagate by seed (self-sows well). Even doubles widely available.

HYDRANGEA
(Hydrangea)
Deciduous shrubs; size varies, but mophead types to 4½ft (1.5m). Flowers late summer/fall, in red, blue, mauve, white. Part or full shade; some hardy to zone 4, others only to zone 7. Remove old flower heads without damaging buds below. Propagate by cuttings. Widely available.

ILEX
(Holly)
Evergreen trees, to 33ft (10m) × 16ft (5m), commonly pruned. Grown for fall berries and foliage, variously shaped and variegated. Sun or shade; hardy to zone 7. Propagate by seed or summer cuttings. Widely available.

IPOMOEA
(Morning glory)
Annual, sometimes perennial, twining climber, indeterminate size, but at least 10ft (3m). Flowers begin when stems are 3ft (1m) high, in blues, magenta, purples, white. Sun; tender. Rich soil, average to moist; plenty of space. Propagate by seed. Widely available.

IRIS
Semievergreen rhizomatous perennial, or bulbs, up to 4–5ft (1.25–1.5m) × indefinite. Flowers in early spring to high summer, in blue, yellow, white. Needs sun; *I. germanica* (German iris) forms, *I. foetidissima* (stinking iris), and a number of bulbous varieties are hardy to zone 4 or 5. All varieties prefer free-draining sandy soil, dry to average moisture. Rhizomatous sorts need dividing every few seasons. Propagate by division or seed. Widely available.

JASMINUM
(Jasmine)
Twining shrubs, often evergreen, sometimes scramblers, indeterminate size. *J. nudiflorum* flowers in winter, others in spring under glass, high summer outdoors; flowers are white, yellow, rarely pink. Sun, part shade; some species hardy to zone 6, others only to zone 10. Needs rich garden soil, allow plenty of space. Propagate by cuttings. Widely available.

LATHYRUS ODORATA
(Sweet pea)
Climbing annual, up to 10ft (3m). Flowers appear in summer or late summer, in purples, blues, pinks, often mixed. Sun; generally dies after flowering, though young plants can be overwintered outdoors. Rich soil, average to moist; deadhead regularly to prolong flowering season. Propagate by seed. Widely available.

LAURUS
(Bay laurel, sweet bay)
Suckering evergreen tree, usually pruned, often as topiary. Height to 33ft (10m), width to 23ft (7m). Flowers small, yellowish, appear in spring; foliage usually green, narrow in 'Angustifolia,' yellow in 'Aurea.' Sun; hardy to zone 8. Propagate by cuttings, seed. Widely available.

LAVANDULA
(Lavender)
Short-lived evergreen bush, dwarf sorts to 8in (20cm), large ones to 3ft (1m) in both height and spread. Flowers in summer, in mauves, white, pale pink. Sun; hardiness varies: *L.* 'Munstead' hardy to zone 5, *L. angustifolia* to zone 6, *L. dentata* to zone 8. Propagate by spring cuttings, seed. Many varieties widely available.

LILIUM
(Lily)
Bulbous perennials, up to 6½ft (2m) high. Flowers in early to high summer, in white, purple, red, yellow, often spotted. Sun or part shade; most species hardy to zone 4, some less hardy. Average to peaty soil, average moisture; watch for tobacco mosaic virus. Propagate by seed or bulb scales treated as cuttings. Most varieties listed widely available. *L. bulbiferum* may be grown from seed.

LONICERA
(Honeysuckle)
Twining shrubs, some evergreen, indeterminate size. Flowers in late spring to early summer, in cream, yellow, red, or purple outer surfaces. Tolerate shade, but best in sun; some hardy to zone 4, others only to zone 9. Watch for aphids. Propagate by layering. Widely available.

LUNARIA
(Honesty, money plant)
Biennial, or perennial in *L. rediviva*, 3ft (1m) × 2ft (60cm). Flowers appear in late spring/early summer, in *L. annua* leaves may be plain or variegated. Sun or shade; hardy to zone 6. Often self-sows. Widely available.

LUPINUS
(Lupine, tree lupine)
Annuals, herbaceous perennials, rarely subshrubs, up to 4½ft (1.5m) × indeterminate. Flowers in summer, in blues, mauves, yellows, white. Sun or part shade; hardy to zone 5. Tree lupines are short-lived, so save a few seeds. Propagate by seed, or division in perennial varieties. Widely available.

LYCOPERSICON
(Tomato)
Herbaceous, often treated as annual, but perennial in tropics, weak shrubs to 6½ft (2m), or rampant climbers. Flowers within a few months of sowing, fruits variously shaped, in reds, yellows, or white. Sun or partial shade; tender. Rich garden soil, average to moist, under glass. Tap open flowers to insure pollination. Old 'Marmande' types widely available.

MALUS
(Apple)
Deciduous tree, varies with variety and the stock used for grafting, but can be up to 30ft (10m) high and across. Mostly pruned to size. Flowers spring, fruit ripening in fall to late winter, in greens, yellows, reds, russets. Sun; most hardy to zone 5. Specialized forms need careful pruning in winter and summer. Propagate by grafting. Many varieties widely available.

MENTHA
(Mint)
Mostly rampant herbaceous perennials, up to 3ft (1m) high × indefinite. Flowers late summer, but grown for their aromatic and tasty leaves; these are variously frilled, plain (often bright) green, or variegated. Sun or shade; many hardy to zone 4. Best grown in pots so that roots can't invade the whole garden. Propagate by division or seed. Widely available.

MESPILUS
(Medlar)
Small deciduous tree, reaching 16ft (5m) × 13ft (4m). Flowers in spring, fruit ripens late fall; flowers incream, fruit warm brown, leaves bronze. Sun or part shade; hardy to zone 6. Propagate either by grafting or by seed. Widely available.

NARCISSUS
(Narcissus, daffodil)
Clump-forming bulbs, up to 1ft 4in (40cm) × indeterminate spread. Flowers appear late winter into early summer, in yellows, cream, rarely white. Sun or partial shade; most hardy to zone 3. Average to well-drained soil, average moisture. Clumps eventually stop flowering; divide every few seasons. Propagate by division. Widely available.

NICOTIANA
(Flowering tobacco)
Often perennial in warmth, generally treated as annuals, up to 6½ft (2m) × 3ft (1m). Flowers summer and into fall (winter under glass), in white, pinks, mauves, green. Sun; hardy to zone 9 or 10. Rich garden soil, average to moist; aphids can be a nuisance. Propagate by seed. Widely available.

OCIMUM
(Basil)
Mostly annuals, some perennials, height and spread 1ft (30cm). Flowers late summer, leaves greens or purple, flat, wrinkled. Sun; tender. Rich garden soil best, average to moist; grow fast to avoid plants running to flower. Will not grow outdoors in northern parts. Widely available.

OENOTHERA
(Evening primrose)
O. biennis usually treated as annual, to 3ft (1m) × 1½ft (45cm); *O. acaulis* perennial, 1ft (30cm) × 1ft (30cm). Flowers summer, in yellows, sometimes white or apricot. Sun or part shade; hardy to zone 5. Widely available.

OMPHALODES
Creeping herbaceous perennial (*O. cappadocicum* evergreen), height to 8in (20cm), spread indefinite. Flowers in spring, blue, rarely white. Part to full shade; hardy to zone 5. Propagate by division. WA04, W-CA15.

ORIGANUM
(Marjoram, oregano)
Mostly herbaceous perennial, 1½ft (45cm) × 3ft (1m). Flowers summer, leaves greens, yellows, variegated. Sun or part shade; hardiness varieties: some varieties hardy to zone 5, others only to zone 9. Propagate by division or seed. Widely available.

PAEONIA
(Peony)
Herbaceous perennials or shrubs, herbaceous sorts to 2½ft (75cm) × 3ft (1m); shrubs to height and spread of 6½ft (2m). Flowers in early to high summer, in white, pinks, reds, yellow. Part shade; most hardy to zone 4. Propagate named herbaceous sorts by division, rarely seed. Shrubs by seed or grafting. Widely available.

PAPAVER
(Poppy)
Annuals, biennials, or herbaceous perennials, sometimes freespreading, up to 3ft (1m) × (in *P. orientale*) indeterminate. Flowers in early summer, in white, scarlets, pinks, purples. Sun; perennials generally hardy to zone 4. Rich garden soil; divide *P. orientale* every few seasons. Propagate herbaceous sorts by root cuttings, others by seed. Widely available.

PELARGONIUM
(Geranium, scented geranium, zonal geranium)
Mostly low shrubs, usually evergreen (deciduous in bad conditions), size varies with variety, but up to 6½ft (2m), more with support. Flowers in all growing seasons, in white, pink, reds, mauves; foliage very variable. Sun, part shade; most hardy only to zone 10. Needs frequent pruning to keep in shape. Propagate by cuttings, seed. Widely available.

PETROSELINUM
(Parsley)
Herbaceous biennial, mostly grown as annual; before flowering, reaches a height and spread of 1ft (30cm). Flowers in second spring, leaves flat or frilled. Sun, part shade. In cold gardens, it is wise to insure their survival by overwintering a few plants indoors. Widely available.

PHASEOLUS
(Scarlet runner bean, French string bean, lima bean)
Climbing or bushy annual (perennial in warm climates), up to 16ft (5m) × indeterminate spread. Flowers in summer or fall; will crop to first frosts, in white, pink, tomato red; pods in greens, yellows, purples, or variously streaked and splashed in these colors. Best in sheltered, sunny places; not hardy. Prefers rich soil, average to wet. Sow seed outdoors a few weeks after last frost. Widely available.

PHILADELPHUS
(Mock orange)
Deciduous bushes, to 15ft (5m) × 10ft (3m). Flowers appear in early summer, in white. Sun, partial shade; most hardy to zone 5. Prune after flowering. Propagate by cuttings. Widely available.

PISUM
(Pea)
Annual, bushy or climbing, size varies with variety, climbers to 10ft (3m). Pods mostly greens, sometimes purple. Sun or partial shade; most sorts not hardy. Does best in rich garden soil. Pick over crop every few days to keep the plants producing. Widely available.

PRIMULA
(Primrose, primula, auricula)
Herbaceous perennials, auriculas evergreen, those listed up to 1ft (30cm) × indeterminate. Flowers late winter to early summer, white, yellows, reds, rarely green. Part shade or full shade; hardiness varies greatly, but most hardy to zone 5. Border soil or leafmold, average moisture for auriculas, average to moist for others. Primroses need dividing each season after flowering. Propagate all varieties either by seed (sow in fall outdoors in pots) or division. Most species widely available.

PRUNUS AVIUM
(Cherry)
Eventually becomes a large deciduous tree, reaching a height of up to 45ft (14m). Flowers spring, fruiting in early to high summer, flowers white, fall foliage golden or scarlet. Sun or part shade (morellos will fruit in full shade); hardy to zone 4. Propagate by grafting. Widely available.

PRUNUS DOMESTICA
(Plum)
Deciduous trees, can reach 33ft (10m), but often pruned smaller. Flowers spring, sometimes early; fruits summer; fall foliage sometimes bronze or yellow. Sun; generally hardy to zone 5. In cool gardens, choice sorts, especially gages, are best trained against a wall. Propagate by grafting, or by seed in myrobalans and some plums and gages. Widely available.

PRUNUS PERSICA
(Peach)
Small tree, can be grown potted, up to 15ft (5m) × 10ft (3m). Flowers early spring (watch for frost damage), fruits midsummer to early fall, in pink, creams, reds, yellows. Shelter, full sun; hardy to zone 5. Often flowers before bees emerge, so hand pollinate. Propagate by grafting, though Pèche de vigne and nectarines come true from seed. Widely available.

PYRUS
(Pear)
Often large, long-lived, deciduous trees, up to 40ft (12m) × 25ft (7m). Flowers spring, fruit ripening summer to late winter. Fall foliage sometimes butter yellow. Sun; hardiness varies with species, but generally hardy to zone 5. Easily pruned to size or to make espaliers or fans. Propagate by grafting. Widely available.

RANUNCULUS
Corms or herbaceous perennials, up to 3ft (1m) × indefinite. Flowers in early summer, in white, scarlet, yellow. Sun or part shade; cormous sorts lifted in summer, others hardy to zone 5. Perennials need dividing every few seasons. Propagate by seed or division. Corms of many varieties widely available.

RIBES X UVA-CRISPA VAR. RECLINATUM
(Gooseberry)
Deciduous, spiny bushes, height and spread to 6½ft (2m). Flowers spring, fruiting early to midsummer; fruits in black, greens, yellows, smooth or prickly. Sun or partial shade; hardy to zone 3. Can be pruned as standards etc.; watch for sawfly. Propagate by cuttings, seed. Difficult to find in U.S.

ROSA
(Rose)
Climbing roses: deciduous or semi-evergreen, usually thorny scramblers, size varies with variety, but can get to 30ft (10m) or more. Flowers in early summer, sometimes late summer, in white, yellows, pinks. Most need sun, though some will also flower well in shade; generally hardy to zone 4. For antique varieties, pruning is not vital; instead, thin out whole plant. Propagate by cuttings.

Shrub roses: shrubs, sometimes lax, size varies with variety, but up to 8ft (2.5m) × 6½ft (2m). Flowers in late mid- to high summer for most single-season varieties, in white, rose pinks, purples. Best in full sun, partial shade possible; fully hardy, though occasional losses below 14°F (-10°C). Rich soil best, average moisture.

Pruning is important — shrub and species roses should be pruned lightly, since they flower on wood two years old or more. Renewal prune every year by cutting out some of the oldest stems right down to the base, allowing space for strong new shoots to grow. With varieties that flower only once, prune after flowering; with those that repeat-flower, do it in fall. Over a four-year period, the whole bush should have been recycled. Alternatively, leave the bush as a tangle, eventually using a hedge-trimmer to cut the whole thing down to about 6in (15cm) and then letting it regrow. Propagate by cuttings in late summer or early fall, though use seed for new varieties.

Many old gardens still have ancient roses in them, often unnamed. Remember that these may come easily from cuttings. If you find something you like, don't ignore it because it is now anonymous.

Other good varieties worth seeking out are: 'Ville de Bruxelles' — CA03, CAN18, NH01; 'Madame Legras de St. Germaine' — CA03, CA21, CAN18, NH01, OH12; 'Tuscany Superb' (gallica-type rose) — CA21, CAN18, MN36, NH01; 'Robert le Diable' (provence-type rose) — CAN18,

NH01; 'Pompon Blanc Parfait' (damask-type rose) — CA18, ME06, NH01, W-CO03.

ROSMARINUS

(Rosemary)

Evergreen bush, upright sorts to height and spread of 4½ft (1.5m). Flowers in early summer in purples, white, pale pink. Sun, part shade; hardy to zone 7. Propagate either by cuttings or by seed. Widely available.

SANGUINARIA

(Bloodroot)

Creeping herbaceous perennial, 8in (20cm) × indefinite. Flowers early spring, white. Part to full shade; hardy to zone 3. Good garden soil, average moisture. Propagate by division. Widely available.

SANTOLINA

(Lavender cotton)

Low evergreen bushes, height and spread 3ft (1m). Flowers in summer, in cream or yellow, usually removed. Sun; hardiness varies with variety, from zone 6–9. Needs average to poor soil, average to dry. Propagate by cuttings. Widely available.

SCILLA

(Squill)

Bulb, up to 1ft 4in (40cm). Flowers spring, blue, rarely shades of purple, white. Sun; all listed varieties hardy to zone 8. Propagate by division, seed. Widely available.

SILENE

(Catchfly, campion)

Clump-forming herbaceous perennials, up to 1½ft (45cm) × indeterminate. Flowers in late spring/early summer, in white, rose pinks. Part shade; hardiness varies, from zone 3–5. Watch for aphids on flowering stems. Propagate double sorts by division, single ones by seed. Try also the double form *S. maritima*. Most species widely available.

SKIMMIA

Evergreen bushes, 3ft (1m) × 4½ft (1.5m). Flowers spring, though often grown for fall berries, in red, pinkish, or white. Partial or full shade; hardy to zone 7. Male and female plants needed for berries. Propagate by seed or summer cuttings. Widely available.

SPINACIA

(Spinach)

Leafy annual, size varies with variety, but height and spread to 1ft 4in (40cm). Leaves smooth, sometimes savoyed. Sun or shade; tender. Rich, moist garden soil — do not let it dry out; sow small amounts through growing period. Widely available.

TAXUS

(Yew)

Left unpruned, makes a small tree 33–50ft (10–15m) × 15–33ft (5–10m). Flowers insignificant, foliage dark green to brownish yellow. Tolerates full shade; hardy to zone 5 or 6 depending on species. Propagate by cuttings, seed. Widely available.

THYMUS

(Thyme)

Creeping perennials or small shrublets, size varies with species, but height and spread to 1ft (30cm). Flowers summer, mauve or white, leaves various greens, yellows, variegated. Sun; hardiness varies from zone 4–6. Average to poor soil, average to dry. Propagate by cuttings, seed. Widely available.

TROPAEOLUM

(Nasturtium)

Climbers, often treated as annuals, to around 10ft (3m). Flowers summer, early fall, in oranges, scarlet, reds, yellow. Sun or part shade; many tender, though *T. speciosum* and some other tuberous sorts hardy to zone 7. Propagate by seed. Widely available.

TULIPA

(Tulip)

Bulbs, up to 2ft (60cm). Flowers in winter if forced, in spring to early summer outdoors, in every color except blue. Sun, except *T. sylvestris* which needs part shade; most are hardy to zone 3 in dryish soils, but are commonly lifted in summer. *T. acuminata* and *T. clusiana* best lifted and dried in winter. Propagate by division; florists' sorts were grown from seed. Widely available.

VERATRUM

(Veratrum, false hellebore)

Herbaceous perennials, up to 6½ft (2m) × 3ft (1m). Flowers in summer, in green, white, blackish. Part to full shade; hardy to zone 7. Rich soil gives best foliage, average to moist soil; slugs can ruin foliage. Propagate by seed (sow outdoors in fall) or division. Widely available.

VERBENA

Sprawling evergreen subshrub, size varies with variety, but often to 1½ft (45cm) × indeterminate. Flowers all growing season, in mauves, reds, white, often with colored eye. Sun, part shade; some hardy to zone 4, others only to zone 9 or grown as annuals. Feed well, prune when messy. Propagate by cuttings, seed. Widely available.

VIBURNUM

Deciduous or evergreen shrubs, *V. opulus* to 22ft (7m) × 10ft (3m); *V. tinus* to 16ft (5m) × 10ft (3m). *V. opulus* flowers in summer; *V. tinus* fall to spring, in white, but both grown for scarlet berries. Partial to full shade; *V. opulus* hardy to zone 4, *V. tinus* to zone 8. Any soil, average moisture, put several plants together for good crops of berries. Propagate by seed, cuttings. Widely available.

VINCA

Low, sprawling evergreens, up to 1ft (30cm) × indefinite. Flowers in late spring, leaves may be variously variegated, flowers blues, purples. Sun or shade; *V. minor* hardy to zone 4, others to zone 7 or 8. Watch that they do not shade out nearby plants. Propagate by division, cuttings. Widely available.

VIOLA

Mat-forming perennials, 1ft (30cm) × indeterminate. Flowers in late spring to fall, depending on species and variety, in white, yellow, blues, purples, reds. *V. odorota* best in shade, others in sun or partial shade; *V. odorota* hardy to zone 4. Good rich moist soil. Spray regularly. Propagate by division, cuttings, seed. Widely available;.

VITIS

Rampant deciduous climber, virtually indefinite height and spread, usually hard pruned. Lightly fragrant flowers in spring, fruit ripening summer or fall; fall foliage sometimes bronzes, fruits green, gold, purples. Sun; most hardy to zone 5. Vegetation needs constant thinning during summer; bunches of fruit need thinning as each berry grows. Propagate by cuttings. Widely available.

KEY TO SOURCES

CA03: Roses of Yesterday and Today, ($3.00), 802 Brown's Valley Rd., Watsonville, CA 95076-0398
408-724-3537

CA21: Greenmantle Nursery ($3.00), 3010 Ettersburg Rd., Garberville, CA 95442
707-986-7504

CAN18: Pickering Nurseries Inc. ($4.00) 670 Kingston Rd., Pickering, Ontario L1V 11A6, Canada
905-839-2111

CT01: Select Seeds ($3.00), 81 Stickney Hill Rd., Union, CT 06076-4617
203-684-9310

ME06: Royal River Roses ($3.00) 70 New Gloucester Rd., North Yarmouth, ME 04097
207-829-5830

MI01: Far North Gardens ($2.00) 16785 Harrison, Livonia MI 48154
810-486-4203

MN36: Orion Farm, 4186 75th St. SW, Waverly, MN 55390
800-558-4280

NH01: Lowe's Own-root Roses ($2.00), 6 Sheffield Rd., Nashau, NH 03062
603-888-2214

OH12 : Historical Roses (SASE), 1657 West Jackson St., Painsville, OH 44077
216-357-7270

WA04 : Collector's Nursery, 16804 NE 102nd Ave., Battle Ground, WA 98604
360-574-2832

Wholesale Nurseries

W-CA15: Native Sons Wholesale Nursery, 397 West El Campo Rd., Arroyo Grande, CA 93420
805-481-5996

W-CO03: High Country Rosarium, 1717 Downing St., Denver, CO 80218
303-832-4026

W-MI02: Far North Gardens ($2.00) 15785 Harrison, Livonia, MI 48154
810-486-4203

Temperature zone map

Selecting plants that are adapted to your climate is an important secret in successful gardening. Your best tool for identifying such plants is the hardiness zone map shown here. Developed by the United States Department of Agriculture (USDA), it divides Canada and the United States into 11 zones based on average minimum winter temperatures. Most plant descriptions — in catalogs and on plant labels, for example — indicate the zones in which a specific plant will thrive. If you buy plants that are recommended for the zone you are in, you can be reasonably confident that they will be adapted to your local climate.

However, plant hardiness can also be influenced by local conditions. A garden located at the top of a 1,000-ft (300-m) mountain, for example, will usually have temperatures several degrees colder than one in the surrounding plains. And because water collects heat from the sun, a garden located on the shore of a pond will be both warmer in winter and cooler in summer than a garden just down the road. You may have to adjust your use of the zone map accordingly.

For further information about plant hardiness in your area, contact your local Cooperative Extension or Agricultural Service.

		Fahrenheit	Celsius
Zone 1		below -50°	below -46°
Zone 2		-50° to -40°	-46° to -40°
Zone 3		-40° to -30°	-40° to -34°
Zone 4		-30° to -20°	-34° to -29°
Zone 5		-20° to -10°	-29° to -23°
Zone 6		-10° to 0°	-23° to -18°
Zone 7		0° to 10°	-18° to -12°
Zone 8		10° to 20°	-12° to -7°
Zone 9		20° to 30°	-7° to -1°
Zone 10		30° to 40°	-1° to 4°
Zone 11		above 40°	above 4°

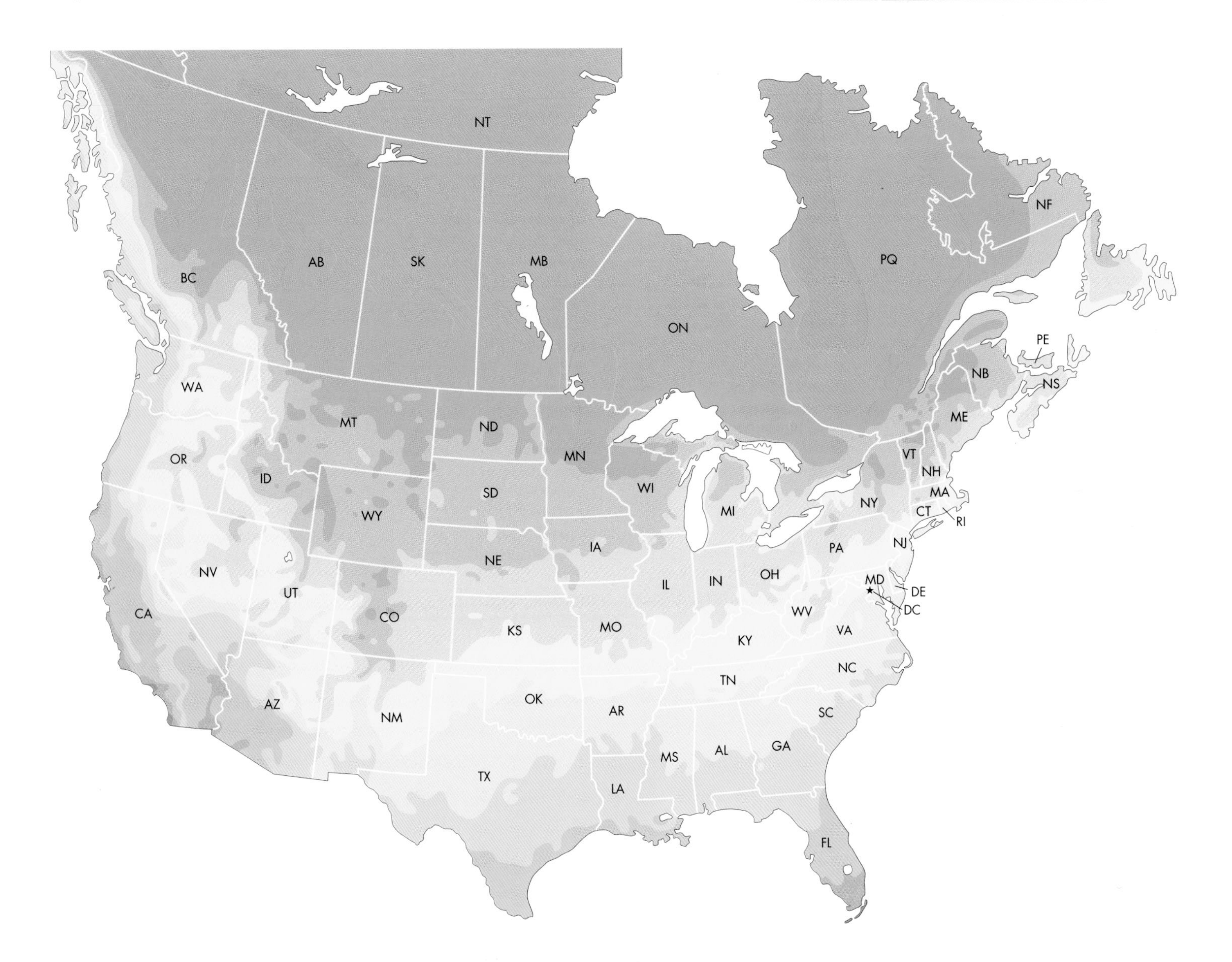

Index

AUTHOR'S ACKNOWLEDGEMENTS

I should like to thank a number of people who have been helpful in the writing of this book. One to whom I am especially grateful is Rand Lee, horticultural writer and editor, from Santa Fe, New Mexico. Founder and current president of the American Dianthus Society and, with Nancy and Ira McDonald, co-creator of *The American Cottage Gardener*, a literary gardening quarterly devoted to interpreting the cottage-gardening style for North American climates and cultivars, he helped with many antique queries. In the UK, I thank Ray Warner for his lists of antique bulb varieties and of vegetables that remain easily available to the gardener (which he has made more available still). I would also like to thank Carla Oldenburger-Ebbers, librarian at the University of Wageningen, Netherlands, for drawing my attention to some early sources of work about topiary.

Thanks go also to the owners of the gardens featured in book, particularly to David Beeson, for much information about meadows and meadow plants; to Roger Warner for help about the history and planting of his conservatory; to Dr. Leslie and Mrs. Cox for telling me about their decorative kitchen gardens (and for a plant of the red-leafed sorrel), as well as to Mr. Jeffrey, gardener at the "topiary garden."

ACKNOWLEDGEMENTS

1 Andrew Lawson; **2–3** Saxon Holt; **5** Bridgeman Art Library/Victoria & Albert Museum, London; **6** Bridgeman Art Library/Central Saint Martins College of Art & Design, London; **7** Bridgeman Art Library/Lindley Library, RHS, London; **8** Bridgeman Art Library/Lindley Library, RHS, London; **9** Bridgeman Art Library/Private Collection; **10** Bridgeman Art Library/Private Collection; **11** Bridgeman Art Library/Private Collection; **12** Bridgeman Art Library/Lindley Library, RHS, London; **13** Bridgeman Art Library/Linnean Society, London; **14** Bridgeman Art Library/Jessica Tcherepnine (living artist), Private Collection; **16** Bridgeman Art Library/Victoria & Albert Museum, London; **17** left Bridgeman Art Library/Private Collection; **17** right Bridgeman Art Library/ Private Collection; **18** S & O Mathews (12 Rozelle Close, Littleton, Hampshire); **19** Bridgeman Art Library/Natural History Museum, London; **20** Bridgeman Art Library/ Mallett Gallery; **21** Bridgeman Art Library/Lindley Library, RHS, London; **22** Andrew Lawson; **23** above S & O Mathews; **23** below Sunniva Harte; 24 Andrew Lawson; **25** above left John Glover; **25** above right S & O Mathews; **26** S & O Mathews; **27** Clive Nichols; **28** left S & O Mathews; **28** right Saxon Holt; **29** left S & O Mathews; **29** right John Glover; **30** above Jerry Harpur; **30–1** S & O Mathews; **30** below Harry Smith Collection; **31** John Glover; **32** Andrew Lawson; **33** above Sunniva Harte; **33** below Andrew Lawson; **34–5** Roger Foley (Colonial Williamsburg); **34** left John Glover; **35** right Andrew Lawson; **36** Sue Snell (Courtesy of The Charleston Trust, Sussex); **37** Andrew Lawson; 38 S & O Mathews; **39** Howard Rice; **40** left Andrew Lawson; **40** right S & O Mathews; **41** John Glover; **42** Photos Horticultural; **43** left John Glover; **43** right S & O Mathews (Little Court, Hampshire); **44** above Howard Rice; **44** below Ken Druse; **45** Jerry Harpur; **46** above S & O Mathews; **46** below Richard Felber (Pam Kay); **47** above left Richard Felber (Pam Kay); **47** above right Marianne Majerus; **47** below S & O Mathews; **48** above S & O Mathews; **48** left & right Richard Felber (Pam Kay); **49** Richard Felber (Pam Kay); **50** S & O Mathews (Stitches Farm House, Sussex); **51** Bridgeman Art Library/Private Collection; **52** Bridgeman Art Library/ Christopher Wood Gallery, London; **56** Andrew Lawson; **57** left Andrew Lawson; **57** right Clive Nichols; **58** above Andrew Lawson; **58** below John Glover; **58–9** S & O Mathews; **60–1** Photos Horticultural; **60** left Clive Nichols; **61** Andrew Lawson; **62** left Andrew Lawson; **63** S & O Mathews; **64** Andrew Lawson; **65** Clive Nichols; **66–7** John Glover; **68** left John Glover; **68** right Andrew Lawson; **69** above S & O Mathews; **69** below Andrew Lawson; **70–1** Andrew Lawson; **70** above left Garden Matters; **70** below left Andrew Lawson; **71** Andrew Lawson; **74** above S & O Mathews; **76** Andrew Lawson (Levens Hall); **78** Bridgeman Art Library/John Spink Fine Watercolours, London; **80** John Glover (Toad Hall, Berkshire); **81** S & O Mathews; **82** above Jerry Harpur (La Casella, Alpes-Maritimes, France); **82** below John Glover (Hever Castle, Kent); **83** Andrew Lawson (Buckland Abbey, Devon, courtesy of The National Trust); **84** Jerry Harpur (La Casella, Alpes-Maritimes, France); **85** above right John Glover (Queen's Garden, Kew Gardens, Surrey); **85** centre right John Glover (RHS Wisley, Surrey); **85** below Jerry Harpur (Chevening, Kent); **86–9** Fritz von der Schulenburg/The Interior Archive (John Stefanidis); **90** The Garden Picture Library/Mayer/Le Scanff; **91** Bridgeman Art Library/Private Collection; **92** Bridgeman Art Library/Chris Beetles Ltd, London; **93** E.T. Archive; **94** Andrew Lawson; **96–7** The Garden Picture Library/ Marijke Heuff; **96** left Harry Smith Collection; **97** right Andrew Lawson; **98–9** The Garden Picture Library/John Glover; **98** above left Saxon Holt; **98** below left Harry Smith Collection; **99** above right Andrew Lawson; **99** below right The Garden Picture Library/Sunniva Harte; **100** David Cavagnaro; **101** below left Photos Horticultural; **101** above left Jacqui Hurst; **101** right Andrew Lawson; **102** Andrew Lawson; **103** Andrew Lawson; **104** left The Garden Picture Library/David Russell; **104** right Andrew Lawson; **105** Andrew Lawson; **106–9** Sunniva Harte (Charleston Manor, Sussex); **110** Richard Felber (Robert Jakob & David White); **111** Bridgeman Art Library/Private Collection; **112** Sotheby's, London; **113** Bridgeman Art Library/Lindley Library, RHS, London; **115** above right The Garden Picture Library /Howard Rice; **115** below right Andrew Lawson; **116** left The Garden Picture Library/John Glover; **116** centre Photos Horticultural; **116** right John Glover; **117** left Jerry Harpur (Drummond Castle); **117** centre Photos Horticultural; **117** right The Garden Picture Library/Neil Holmes; **118** Sue Snell (Courtesy of The Charleston Trust, Sussex); **119** above Ken Druse; **119** below S & O Mathews; **120** above left Christian Sarramon; **120** below left Christian Sarramon; **120** right John Glover; **121** The Garden Picture Library/Lamontagne; **122–3** Saxon Holt; **122** left Juliette Wade; **122** right Sunniva Harte; **123** right Photos Horticultural; **124** Marijke Heuff (Huis Bingerden, Holland); **125** Clive Nichols (Cerney House Garden, Gloucestershire); **126** Andrew Lawson; **127** above Roger Foley (Colonial Williamsburg); **127** below centre Photos Horticultural; **127** below right John Glover; **127** right S & O Mathews; **127** below left S & O Mathews; **129** right John Glover; **131** above Roger Foley (Stratford Hall); **132** above S & O Mathews; **133** right The Garden Picture Library/Vaughan Fleming; **134** The Garden Picture Library/ Linda Burgess; **135** Bridgeman Art Library/ Fitzwilliam Museum, University of Cambridge; **136** Bridgeman Art Library/Natural History Museum, London; **137** Bridgeman Art Library/Private Collection; **138** below Jacqui Hurst (Lady Gibberd's Garden, Harlow, Essex); **138** above Andrew Lawson; **139** Howard Rice; **140** above Brigitte Perdereau (Erwan Tymen); **140** below S & O Mathews; **141** above Clive Nichols (Red Gables, Worcestershire); **141** below S & O Mathews; **142–3** S & O Mathews; **143** right Andrew Lawson; **145** Andrew Lawson; **146** above left Marion Nickig; **146** below left Marijke Heuff (Walda Pairon, Belgium); **148** left John Glover; **149** above S & O Mathews.

The following photographs were specially taken for Conran Octopus: Marianne Majerus courtesy of Dr. & Mrs. Cox, Woodpeckers, Warwickshire: **114, 130–1, 130** above, **131** below, **132** below, **133** left & centre. Marianne Majerus: **146** below, **147, 148** right, **149** below. Stephen Robson courtesy of Annette & David Beeson, Forest Edge, Hampshire: **54–5, 72, 73, 74–5, 128, 129** left.